Consumers, Meat and Animal Products

This book addresses the production practices employed in the production of food animals and animal products that enable marketers to sell a variety of products to meet consumer demand.

Food animal production practices have come under increased scrutiny by consumers who object to inputs and practices. The industry has been a proponent of using technologies to reduce production costs, resulting in lower-priced meat and animal food products, and now consumers are starting to look at other objectives. This book considers the key issues of concern to consumers, including the treatment of animals, the use of antibiotics, feed additives and hormones, and how these are monitored, regulated, and communicated to consumers. It also reviews labeling and information provided to consumers, including organic, genetic engineering, welfare standards, and place of origin. While the main focus is on the United States, there are descriptions of European practices and legislation.

Overall, it aims to provide an objective and balanced appraisal, which will be of interest to advanced students and researchers in agricultural, food and environmental economics, law and policy, and animal production and welfare. It will also be very useful for early career professionals in the food and agricultural sectors.

Terence J. Centner is Professor of Practice in the Institute of Agriculture and Natural Resources and Adjunct Professor in the College of Law at the University of Nebraska, Lincoln, USA. He teaches law courses and engages in research involving issues confronting agriculture and the environment. He is also Professor Emeritus at the University of Georgia, USA. Professor Centner has served as an Alexander von Humboldt Stiftung research fellow at the University of Göttingen, Germany; a Fulbright Scholar at the University of Mannheim, Germany; and a Fulbright–Scotland visiting professor at the University of Aberdeen, UK. He has lectured in 50 countries around the world.

Other books in the Earthscan Food and Agriculture Series

Contested Sustainability Discourses in the Agrifood System
Edited by Douglas H. Constance, Jason Konefal and Maki Hatanaka

The Financialization of Agri-Food Systems
Contested Transformations
Edited by Hilde Bjørkhaug, André Magnan and Geoffrey Lawrence

Redesigning the Global Seed Commons
Law and Policy for Agrobiodiversity and Food Security
Christine Frison

Organic Food and Farming in China
Top-down and Bottom-up Ecological Initiatives
Steffanie Scott, Zhenzhong Si, Theresa Schumilas and Aijuan Chen

Farming, Food and Nature
A Sustainable Future for Animals, People and the Environment
Edited by Joyce D'Silva and Carol McKenna

Governing Sustainable Seafood
Peter Oosterveer and Simon Bush

Farming Systems and Food Security in Africa
Priorities for Science and Policy Under Global Change
*Edited by John Dixon, Dennis P. Garrity, Jean-Marc Boffa,
Timothy Olalekan Williams, Tilahun Amede with Christopher Auricht,
Rosemary Lott and George Mburathi*

Consumers, Meat and Animal Products
Policies, Regulations and Marketing
Terence J. Centner

For further details please visit the series page on the Routledge website: www.
routledge.com/books/series/ECEFA/

Consumers, Meat and Animal Products
Policies, Regulations and Marketing

Terence J. Centner

LONDON AND NEW YORK

from Routledge

First published 2019
by Routledge
2 Park Square, Milton Park, Abingdon, Oxon OX14 4RN

and by Routledge
52 Vanderbilt Avenue, New York, NY 10017

First issued in paperback 2020

Routledge is an imprint of the Taylor & Francis Group, an informa business

British Library Cataloguing-in-Publication Data
A catalogue record for this book is available from the British Library

Library of Congress Cataloging-in-Publication Data
Names: Centner, Terence J., author.
Title: Consumers, meat and animal products : policies, regulations
    and marketing / Terence J. Centner.
Description: Milton Park, Abingdon, Oxon ; New York,
    NY : Routledge, 2019. | Series: Earthscan food and agriculture |
    Includes bibliographical references and index.
Identifiers: LCCN 2018054374 (print) | LCCN 2018056309 (ebook) |
    ISBN 9780429430572 (eBook) | ISBN 9781138365797 (hbk) |
    ISBN 9780429430572 (ebk)
Subjects: LCSH: Meat industry and trade. | Animal products. |
    Food animals—Feeding and feeds.
Classification: LCC TS1955 (ebook) | LCC TS1955.C46 2019 (print) |
    DDC 338.1/76—dc23
LC record available at https://lccn.loc.gov/2018054374

ISBN 13: 978-0-367-67143-3 (pbk)
ISBN 13: 978-1-138-36579-7 (hbk)

Typeset in Goudy
by Apex CoVantage, LLC

I dedicate this book to my parents, Harry E. and Mary Ellen Centner. As the owners of a small family farm in western New York state, they introduced me to a lifelong interest in sustaining agricultural production while practicing environmental stewardship.

Contents

Figures

Tables

Preface

The production of food animals and marketing activities connected to meat and animal products are accompanied by concerns and controversies. Consumers often have opinions on various production practices without understanding the tradeoffs of costs, animal welfare, food safety, and free enterprise. Marketing firms have responded to consumer concerns by using labels on food products that identify special attributes. Governments have listened to consumers' safety concerns and have adopted numerous regulations to reduce risks of contaminated food products.

Descriptions of food safety efforts and the meat industry provide a foundation for analyzing America's production of meat and animal products and the many product choices being offered to consumers. Consumers want information on the inputs used in the production of cattle, dairy cows, pigs, chickens, and seafood. Production practices may involve the use of antibiotics, hormones, beta agonist feed additives, and pesticides. Consumers have supported regulations that keep animals healthy and preclude practices that might contribute to unwholesome food products. In addition, consumers are concerned about production practices involving pain and animal suffering. A few consumers also want to know about breeding, cloning, and genetic engineering related to animal production.

Given consumer opposition to selected animal production practices, marketing firms are responding by placing more information on product labels. This information often identifies the absence of objectionable practices and inputs. Although the products may be costlier, consumers are expressing their preferences and purchasing the specialized products. Additional rules for production and marketing practices may be advocated in order to respond to other issues important to American consumers.

Acknowledgments

I deeply appreciate the support of several organizations that have provided me opportunities to learn more about various topics I have investigated, including animal production practices and issues involving the marketing of animal products. As a faculty member in the College of Agricultural and Environmental Sciences at the University of Georgia, I was able to develop research and teaching programs that led to this book. The College also supported my participation in three academic programs in Europe. Under the auspices of the Alexander von Humboldt Foundation (Bonn, Germany), I spent a semester at the University of Göttingen. Under the auspices of the J. William Fulbright Foreign Scholarship Board (US Department of State, Washington, DC), I taught semester-long courses at the University of Mannheim and the University of Aberdeen. During my tenure at Georgia, I was able to attend numerous academic conferences at which I learned other viewpoints about animal production practices and animal welfare.

I would also thank my colleagues and students at the University of Georgia for their support. Numerous coauthors were instrumental in expanding the scope of my research and enhancing the analyses. My students continually amazed me with their inquiries, their novel approaches to analyzing issues, and their desire to learn more. Colleagues at other universities and members of the American Agricultural Law Association also provided their experiences, insights, and observations, which broadened my perspectives. And currently, I am pursuing my teaching and writing projects at the University of Nebraska, and appreciate their support. Finally, above all, I express my special thanks to my family for their support of my travel and academic pursuits.

Abbreviations

BST	bovine somatotropin
Bt	*Bacillus thuringiensis*
CAFO	concentrated animal feeding operation
CDC	US Centers for Disease Control and Prevention
COOL	country of origin labeling
CRISPR	clustered regularly interspaced short palindromic repeats
DNA	deoxyribonucleic acid
EC	European Commission
EFSA	European Food Safety Authority
EID	electronic identification
EPA	Environmental Protection Agency
EU	European Union
FAO	Food and Agriculture Organization
FDA	Food and Drug Administration
FIFRA	Federal Insecticide, Fungicide, and Rodenticide Act
GE	genetically engineered
GMOs	genetically modified organisms
GRAS	generally recognized as safe
FSIS	Food Safety and Inspection Service
HACCP	Hazard Analysis and Critical Control Point
ISO	International Organization for Standardization
kg	kilogram
mg	milligram
MRL	Maximum Residue Limit
NOP	National Organic Program
NPDES	National Pollutant Discharge Elimination System
rBST	recombinant bovine somatotropin
TALENs	transcript activator-like effector nucleases
UN	United Nations
USDA	United States Department of Agriculture
WHO	World Health Organization
WTO	World Trade Organization
μg/kg bw	microgram per kilogram of body weight
ZFNs	zinc finger nucleases

Part I
Safety and the industry

1 Introduction

Key questions to consider

1   What are we looking for when we make food purchases?
2   How does the US reduce risks of contaminated food?
3   How do we guarantee the truthfulness of food attributes?
4   Do we understand information on attributes provided by food labels?

We are accustomed to having choices in deciding what to eat and what groceries to buy. Many of us love the large variety of fresh and prepared food items that are available at local grocery stores as well as those affiliated with specialty, national, and regional firms. Moreover, food in America is generally not expensive. We spend less of our incomes on food than people living in any other country.

With the large variety and low costs, it is not surprising that many of us are willing to pay more for food products touting specialized attributes. In response to consumer demand, food producers and marketers are conveying more information to consumers, as well as providing specialized products. They are using product labels to tell us about their products so we will buy them and are seeking additional features to differentiate their products.

With our wealth of food choices, a visit to the grocery store can be accompanied by so much information that we experience frustration. What do terms like "organic," "no sugar added," and "no added hormones" really mean? Does "organic" mean the food has been grown without synthetics? No, when we read the organic regulations, we learn that dozens of synthetic substances can be used in organic production and may be present in organic food products.

Does "no sugar added" mean a product has less sugar than a comparable product without a similar claim? No, there are natural sugars. Products with added sugar may actually have lower amounts of sugar than products with natural sugar. Are marketers adding hormones to food products? No, a label with "no added hormones" addresses the production of food animals and tells us the animals did not receive added hormones while being raised.

These three examples help demonstrate why many Americans are confused. We do not understand how food is produced, and labeling conventions do not always provide the information we want. Moreover, we sometimes do not comprehend the full meaning of labeled products when making our grocery purchases. As noted by Dr. Louis W. Sullivan, former Secretary of the US Department of Health and Human Services:

> The grocery store has become a Tower of Babel, and consumers need to be linguists, scientists and mind readers to understand the many labels they see. Vital information is missing, and frankly, some unfounded health claims are being made.
>
> (Hilts, 1990)

Why are labels being used if many people cannot understand them? Marketers are attaching confusing and meaningless labels to food products to entice consumers to buy them even though the products may not be special. This can be called "foodwashing." Just as marketers use "greenwashing" to convey various information falsely suggesting environmental friendliness and benefits, marketers use foodwashing to tout abstract and insignificant qualities about food products to justify higher prices.

An example is a "sustainably raised" label on a meat product. There is no scientific or agreed-upon definition for this term. A second example is "farm raised." Other than lab-grown meat, is there any animal product that comes from an animal not raised on a farm?

Many marketers and consumers have also commented on the ambiguous and meaningless term "natural" used on food products, urging the US Food and Drug Administration (FDA) to not permit this term on labels (Hooker et al., 2018). Finally, a "no sugar added" label tells us nothing about the amounts of natural sugars in a product and so is not helpful for consumers attempting to avoid calories.

With the identification of ambiguous and meaningless information that leads consumers to buy a product, goods may be selected without any real benefits. Since the prices of these products are generally higher than those for other products, consumers are not receiving any value for their extra expenditures.

Through a discussion of 18 topics, this book illuminates issues concerning food attributes and foodwashing. Each chapter commences with key questions that are discussed in the subsequent text. Chapters conclude with a few foodwashing facts that were highlighted in the chapter to summarize significant points. Through these questions and facts, readers can learn about the major policies, regulations, and marketing strategies related to meat and animal products. With this information, consumers can discern what aspects of food products are important to them.

Safe food

Our first concern about food products is safety. When purchasing food, we expect the products to be safe to eat. We expect ground beef not to contain *E. coli* O157:H7, a potentially deadly bacterium that can cause dehydration, bloody diarrhea, and abdominal cramps. We reckon that packaged greens will be free from *Cyclospora cayetanesis*, a parasite that causes an intestinal illness.

We assume we can depend on the label information for identifying foods that are known to contain eight allergens – milk, eggs, fish, crustacean shellfish, tree nuts, wheat, peanuts, and soybeans. These allergens can cause severe allergic reactions in some people, so we have rules that foods containing these allergens need to be labeled (US Code of Federal Regulations, 2018, tit. 21). With the information, persons with allergies can avoid these ingredients.

While we could hope that every food producer, retailer, and marketing firm would always provide safe products, we know that lapses occur. Safety measures cost money, and some firms fall short in taking sufficient precaution to keep their products safe. Experiences with past illnesses and deaths from contaminated food products have led federal, state, and local governments to become involved in efforts to reduce the likelihood of contamination that will harm people consuming food products.

The FDA has been authorized to take actions necessary to protect consumers against impure, unsafe, and fraudulently labeled food and drug products. The US Department of Agriculture's (USDA's) Food Safety and Inspection Service is the public health agency responsible for ensuring that the nation's commercial supply of meat, poultry, and egg products are safe and correctly labeled. The federal Centers for Disease Control and Prevention gathers data and investigates foodborne illnesses and outbreaks to provide information that is useful in helping reduce foodborne illnesses.

Many of our state legislatures have decided that our federal efforts are not enough. A majority of states have authorized their departments of public health to develop food safety programs to assure citizens that foods are safe and are not adulterated, misbranded, or falsely advertised. States have initiated inspection programs for food processing plants and food warehouses. State agencies can embargo or recall questionable food, investigate complaints, engage in the destruction of unwholesome food, coordinate foodborne outbreak investigations, and analyze the findings. The agencies can use this information to develop regulations and guidance designed to prevent similar outbreaks in the future. States also help local health departments by training personnel, reviewing programs, and implementing regulations.

What else do we want?

Today, most of us are demanding more than safe food. We want healthy food. We want information on sugar, fats, cholesterol, and vitamins. We want fresh fruits and vegetables and often avoid blemished produce. Turning to animal products, many of us want to avoid products that have a connection to added hormones, feed additives, and antibiotics. For example, we look for milk products that did not come from cows injected with recombinant bovine somatotropin. To secure desired products, we need labels with accurate information.

We also want the animals to be treated humanely. We do not want veal from calves caged in crates or small pens. We look for cage-free eggs so we know the hens were not confined in tiny wire cages. We have a variety of beliefs about

Table 1.1 Food product attributes and consumer responses in seeking or avoiding the attribute

General products Characteristic	Seek or avoid	Animal products Characteristic	Seek or avoid
Organic	Seek	Production practices	Both
GMOs	Avoid	Antibiotics	Avoid
Sugar	Avoid	Feed additives	Avoid
Local food production	Avoid	Cloning	Avoid
Country of origin	Seek	Humane treatment	Seek
Pesticides	Both	Space	Seek
Obesity concerns	Avoid	Chickens	Both
Natural	Avoid	rBST	Avoid
Health concerns	Both	Hormones	Avoid

the importance of animal welfare in our purchasing decisions. To avoid products connected to undesirable inputs or treatment, we need truthful label information that producers did not use the objectionable input.

Our search for healthy food leads many to seek specialized food items identified with labels describing attributes. We are willing to pay for these special characteristics as we feel the extra costs are worth it. Table 1.1 summarizes 18 major attributes consumers are seeking or avoiding in food products. Sometimes, we seek labels denoting the attributes we like, while for some objectionable attributes, we want labels so we can avoid these products.

Simultaneously, there are many people who want low-cost food items that are safe to consume. More than 40 million Americans fend off hunger with federal assistance, and 11 million live in households that experience hunger (Price, 2017). These Americans are willing to accept foods that do not have specialized health and cosmetic characteristics. Thus, there is a demand for low-priced basic foodstuffs as well as for higher-priced specialty products.

Because not everyone wants the same products, our supermarkets offer a variety of products. This variety keeps changing as marketers react to consumer preferences to maintain business and make a profit. Food producers also respond to markets. If consumers are willing to pay more for a specialized product, such as organic meat or cage-free eggs, it will be supplied as long as it is profitable for producers and marketers.

Labeling information to facilitate choices

Labels on food products constitute the interface of consumers' purchase decisions and the functioning of a market for differentiated food products (Tonkin et al., 2015). In the absence of face-to-face encounters when products are sold, labeling provides the communication between producers and consumers. Consumers are dependent on truthful and accurate labels to provide the information required to make choices in selecting products they desire. Information on labels influences

Table 1.2 Summarizing consumer concerns and objectives for labels

Concern	Objective	Examples
Personal health	Consumer well-being	Hormone free; antibiotic free
Animal health	Animal well-being	Cage-free; free range
Animal welfare	Animal well-being	Certified Humane®; no rBST administered
Environment	Less environmental degradation	No pesticides used; grass-fed
Wholesomeness	Feeling of goodness	Organic; no GMO ingredients
Neighbors	Supporting your community	Locally produced
Economic	Supporting selected groups	Locally produced; country of origin

the attitudes and behaviors of consumers as well as future purchase decisions (Verbeke and Ward, 2006).

Our interest in food labels sometimes involves distrust of the food industry in providing healthy food. We may seek a product fortified with additional vitamins. In other cases, we want to express our preference for food items that are different. We want wholesome foods, pesticide-free foods, and environmentally friendly foods. We want foods from certain animals or plants, or we want foods from a certain region. Alternatively, we may want to avoid foods from animals treated poorly that contain unnatural substances or involve genetic engineering.

Table 1.2 draws upon some of the characteristics of Table 1.1 to summarize several aspects of food production identified by product labels. The first column lists major concerns expressed by consumers followed by related objectives. The final column illustrates marketing strategies that set forth label information to respond to consumer concerns.

Through label information on production practices and qualities, marketers entice buyers who desire to express their concerns and objectives by purchasing more expensive specialized products. When shopping at a supermarket, Americans are inundated with choices. By correctly interpreting label information, we can express our concerns and objectives to purchase the items we want.

USDA labeling provisions

The USDA's Food Safety and Inspection Service has enacted various rules on the labeling of meat products (US Code of Federal Regulations, 2018, tit. 9). The agency adopted a general rule that meat food products for human consumption should have nutrition labeling. Unless an exception applies, ground meat products need to have nutrition labels that relate nutrient information to serving size.

Retailers of meat products may offer nutrient content labeling claims concerning products with relation to the Reference Daily Intake or Daily Reference Value (US Code of Federal Regulations, 2018, tit. 9). Claims may include that the item is a "good source" of or "rich in" a nutrient. Thus, a beef product may say it is a "good source of iron."

Meat labels may also identify a product as having protein that helps maintain healthy muscles and bones. A meat product can be labeled with information that it is a good source of zinc, vitamin B12, vitamin B6, niacin, and selenium. Moreover, health and fat claims may also be used on qualifying products (US Code of Federal Regulations, 2018, tit. 9). Thus, the USDA's labeling provisions offer retailers numerous opportunities to communicate with the public about qualities of meat products.

The USDA has proposed a new rule that would amend the nutrition labeling requirements for meat and poultry products (USDA, 2017). The agency realized that its regulations need to better reflect the most recent scientific research and dietary recommendations to assist consumers in maintaining healthy dietary practices. The agency wanted to update the list of nutrients that are required or permitted to be declared, provide updated Daily Reference Values and Reference Daily Intake values based on current dietary recommendations from consensus reports, and amend the labeling requirements for foods represented or purported to be specifically for children and pregnant women. However, with federal agencies being encouraged to reduce the number of regulations, it is unclear whether the beneficial provisions of this proposal will be implemented.

Guarantees for claimed attributes

When we peruse labeled products at a grocery store, we might wonder who is checking to see whether products accurately reflect the ascribed attributes? What if a producer falsely reports an attribute or a dishonest wholesaler or retailer makes an untruthful claim on a product label? How can we be sure that we are buying the desired product?

In considering these questions, we realize that we are dependent on others when buying our food. We depend on producers to accurately account for production practices, and marketers to correctly label products to tell us what we are buying. But we do not simply hope that people will be honest and not make mistakes. Rather, we have asked our government to assist in providing assurances that we are getting what we pay for. Governmental regulations have been adopted to make it illegal to sell goods with false and misleading labels.

For meat products, the Poultry Products Inspection Act precludes misbranding of poultry products, while the Federal Meat Inspection Act prohibits false and misleading labels of non-poultry meat products (US Code, 2012). This means the federal government can initiate actions to prohibit the use of labels that mislead consumers. Due to limited budgets and personnel, the USDA is not always able to monitor and prosecute all mislabeled product claims. Moreover, under both of these laws, state governments and private citizens are precluded from initiating enforcement actions for mislabeled meat and poultry products.

Given limitations in relying on governmental labeling regulations, we have proceeded to recognize accrediting groups to verify for the truthfulness of labeled goods. Publicly authorized and private certification firms are overseeing production and marketing operations to make sure that products are correctly labeled.

For organic products, 80 certifying agents have been approved for overseeing organic production and handling system plans. The organic certifications guarantee that the food item qualifies for the organic label (USDA, 2016). Turning to voluntary efforts, three humane certification programs are available to guarantee that a meat product came from an animal that was treated humanely. We can rely on certifications made by the American Humane Association, Animal Welfare Approved, or Humane Farm Animal Care to select products meeting our objective that the animals providing our food were not mistreated.

Providing information about attributes of food products costs money. In a study comparing the costs of production of organic chicken with regular chicken, research found that the expenses were between 70 and 86 percent more (Cobanoglu et al., 2014). The increased costs were due to the higher cost of feed, labor, certification, and outdoor area maintenance. These costs can be recovered by selling organic meat at higher prices, and for the study, the high price of organic chicken enabled producers to make a profit.

Any administrative regulatory oversight of labeling by federal and state governments is generally borne by taxpayers (Centner, 2017). Even in situations where there are fees, they often are not sufficient to cover all of the administrative costs. For private certification and labeling programs, the costs are borne by producers, sellers, and consumers. This means that specialized products will have higher prices. Consumers with tight budgets and low disposable incomes often choose not to buy these higher-priced products.

Confusion and consequences

With assurances of safe food and products with accurate labels, a subsequent issue is whether we understand the information provided by labels and what this information means for humans, animals, and the environment. Research suggests that many of us do not completely understand or comprehend the meanings of information on food products. An underlying problem is that we are unfamiliar with scientific terms and production practices.

The lack of understanding of science was the topic of a research study about food product labels containing information on non-genetically grown products and deoxyribonucleic acid (DNA). In a survey with a question about non-genetically modified tomatoes, 33 percent of the responding consumers thought these tomatoes did not contain genes (McFadden and Lusk, 2016). Yet all living material has genes. When a survey asked consumers whether they supported a label for food indicating the presence of DNA, 85 percent wanted this information (McFadden and Lusk, 2016). Yet such a label is meaningless because all food products have DNA.

Alternatively, some consumers may not comprehend particular common terminology. For example, what does a label noting "sugar" mean? Does the label refer to sweetness, natural sugars, or added sugar? A study found that one-half of the participants characterized diet soft drinks as sugary, even though diet drinks use sweeteners that are not sugars (Rampersaud et al., 2013). In another experiment, researchers made up a "MUI-Free" label: the label was fictitious. When this fabricated label was placed on food products, consumers felt it described a healthier product even though they had no idea what it meant (Priven et al., 2015).

Consumers may also be confused because they lack sufficient information on how production practices affect the use of resources. Consumers often select organic food to avoid exposure to pesticide residues and synthetic substances (Smith-Spangler et al., 2012). However, organic food may increase risks of contaminated food products, especially greens and meat (Olaimat and Holley, 2012). Research shows that organic meat is more likely to be contaminated with *Campylobacter* (Rosenquist et al., 2013).

Consumers selecting products for health purposes, such as meat products without hormone implants and feed additives, are making choices that have environmental spinoffs. By forgoing these production inputs, more animals are needed to produce the same amounts of meat products (Capper, 2011). This means that greater amounts of feed are needed, more land must be used to produce food for the animals, and more manure will be produced that needs to be disposed of in a manner to avoid environmental degradation.

Scientific studies of food products reveal confusion by researchers. Two researchers from a university food science department were interested in whether consumers associated poor tasting meat products with the use of hormones and antibiotics in animal production. They conducted an experiment that asked participants about the taste of four labeled chicken products, one of which was labeled "No Hormones Added" (Samant and Seo, 2016). However, US law prohibits the use of hormones in chicken production (USDA, 2011). No known labels are being used on chicken products relating to hormones. What a poorly designed and confusing scientific study.

Another issue is whether we should be concerned about irrelevant claims. Do we want to allow label information that serves little or no purpose? For example, should sellers be able to label vegetables as "fat-free" when we know that they do not have fat? Should we allow sellers of chicken products to make the claim that "no hormones were added" when it is illegal under US law to give chickens hormones? To avoid confusion, governmental agencies have adopted regulations that prohibit label information that fails to disclose meaningful information.

Reducing chaos

The goal of this book is to decrease the chaos when interpreting label information on food products and to help readers avoid being foodwashed. This is a difficult task as we have dissimilar knowledge, want disparate information, and may interpret the same information differently. Providing accurate and non-misleading information is therefore challenging. While governmental agencies such as the FDA and the USDA have devised comprehensive regulatory systems to protect consumers, they do not always realize what they are receiving when buying food products.

To respond to the chaos that surrounds food labeling, the book proceeds sequentially through four groupings of topics. The first section starts with food safety, followed by a description of the meat industry. The second group of chapters looks at the production of food animals. The topics include information on the production of large food animals, chickens, and seafood, followed by discussions of the humane treatment of animals and crowded growing conditions.

A third group of chapters discuss inputs being used to enhance the economic production of food animals. The practices include the use of antibiotics, growth hormones, feed additives, pesticides, breeding and cloning, and genetic engineering.

The fourth section examines labeling information and social issues that consumers may consider when purchasing meat and products from animals. Separate chapters examine animal production practices, organic production, locally grown products, animal waste management, and nuisances and product disparagement. These chapters highlight the significance of labels in enabling consumers to select products with desirable characteristics.

An analysis of consumer information on food products suggests that a thoughtful regulatory framework can help ascertain that consumers get products indicated by the label. Due to higher prices for specialty products, as well as human error, products may be falsely labeled. Enforcement mechanisms are needed to help prevent fraud. Consumer confusion also suggests that we also may want to limit what labels say. This can assist in avoiding commonly misunderstood terms and not including information that most consumers find unnecessary.

Foodwashing facts

1 Governmental oversight is needed to reduce lapses in food safety.
2 Consumers want a variety of food products related to health and environmental attributes.
3 Governmental regulations are needed to facilitate the marketing of specialized food products.
4 An unregulated marketplace leads to foodwashing by marketers attempting to garner increased sales.

References

Capper, J.L. 2011. *Sustainability – Beef*. Proceedings of the North American Veterinary Conference, pp. 36–38, Orlando, Florida, USA, January 15–19. Gainesville: The North American Veterinary Conference, CAB Abstracts.

Centner, T.J. 2017. Differentiating animal products based on production technologies and preventing fraud. *Drake Journal of Agricultural Law* 22(2), 267–291.

Cobanoglu, F., et al. 2014. Comparing the profitability of organic and conventional broiler production. *Brazilian Journal of Poultry Science* 16(4), 403–410.

Hilts, P.J. 1990. Congress votes bill on labeling of food and health claims. *New York Times*, sec. A, p. 1, October 25. www.lexisnexis.com/hottopics/lnacademic/?flapID=legal&random=0.6291601523101579.

Hooker, N., et al. 2018. Natural food claims: Industry practices, consumer expectations, and class action lawsuits. *Food & Drug Law Journal* 73, 319–337.

McFadden, B.R., Lusk, J.L. 2016. What consumers don't know about genetically modified food, and how that affects beliefs. *FASEB Journal* 30(9), 3091–3096.

Olaimat, A.N., Holley, R.A. 2012. Factors influencing the microbial safety of fresh produce: A review. *Food Microbiology* 32, 1–19.

Price, T. 2017. Hunger in America. *Congressional Quarterly Researcher* 27(24), 557–580.

Priven, M., et al. 2015. The influence of a factitious free-from food product label on consumer perceptions of healthfulness. *Journal of the Academy of Nutrition and Dietetics* 115(11), 1808–1814.

Rampersaud, G.C., et al. 2013. Knowledge, perceptions, and behaviors of adults concerning nonalcoholic beverages suggest some lack of comprehension related to sugars. *Nutrition Research* 34, 134–142.

Rosenquist, H., et al. 2013. *Campylobacter* contamination and the relative risk of illness from organic broiler meat in comparison with conventional broiler meat. *International Journal of Food Microbiology* 162, 226–230.

Samant, S.S., Seo, H-S. 2016. Quality perception and acceptability of chicken breast meat labeled with sustainability claims vary as a function of consumers' label-understanding level. *Food Quality and Preferences* 49, 151–160.

Smith-Spangler, C., et al. 2012. Are organic foods safer or healthier than conventional alternatives? A systematic review. *Annals of Internal Medicine* 157, 348–366.

Tonkin, E., et al. 2015. Trust in and through labelling – A systematic review and critique. *British Food Journal* 117(1), 318–338.

US Code. 2012. Title 21, Sections 467e, 678.

US Code of Federal Regulations. 2018. Title 9, part 317; title 21, Section 117.135.

US (Department of Agriculture). 2011. *Meat and Poultry Labeling Terms*. Food Safety and Inspection Service. www.fsis.usda.gov/wps/wcm/connect/e2853601-3edb-45d3-90dc-1bef17b7f277/Meat_and_Poultry_Labeling_Terms.pdf?MOD=AJPERES.

USDA 2016. *Accredited Certifying Agents*. www.ams.usda.gov/services/organic-certification/certifying-agents.

USDA. 2017. Revision of the nutrition facts labels for meat and poultry products and updating certain reference amounts customarily consumed. *Federal Register* 82, 6732–6823.

Verbeke, W., Ward, R.W. 2006. Consumer interest in information cues denoting quality, traceability and origin: An application of ordered probit models to beef labels. *Food Quality and Preference* 17, 453–467.

2 Providing safe products and food security

Key questions to consider

1 Why are Americans becoming ill from foodborne illnesses?
2 Should substances be added to food products without governmental oversight?
3 Why is so much of our food not consumed?
4 What might be done to help more Americans have sufficient nutritious food?

Before looking at today's animal production and marketing issues that are most frequently in the news, we need to acknowledge the twin social issues of safe food and food security. While we may feel that food safety is the most important food issue, for millions of people around the world, food security is also a critical issue. People and societies cannot function well if they do not have sufficient, nutritious food.

If a society cannot keep their food supplies safe for human consumption, they will be disadvantaged. Likewise, people lacking nourishment are constrained in the contributions they are able to make to their families and society. Families expending large amounts of their available funds to secure food will lack funds for other purposes, including housing, sanitation, and education. Persons who are food insecure may not be as productive and may have impaired cognitive development. This detracts from economic growth.

Most governments realize that both safety and food security are necessary and take steps to provide food for their populations. Malnutrition places a heavy burden on developing countries and may account for the loss of 2–4 percent of gross domestic product for a country (Berman et al., 2013). Sufficient safe food is necessary to enable citizens to have meaningful and productive lives.

Foodborne illnesses

We want our food to be wholesome and recognize that governmental regulations are very important in maintaining safe food supplies. Food safety means the food items are not contaminated and their condition will not contribute to illness or adverse medical consequences. As a nation, the United States and other developed nations are doing a good job in providing safe foodstuffs for citizens. Due to an extensive range of legislative and regulatory requirements, unhealthy conditions are minimized and food products entering the market are usually safe.

The elimination of most unsafe food products has curtailed diseases and illnesses, reduced absenteeism from work, and lowered medical costs. Yet, an estimated 48 million cases of foodborne illnesses adversely affect Americans every year (FDA, 2016). Moreover, 128,000 hospitalizations and 3,000 deaths per year in our country can be attributed to unsafe food supplies (FDA, 2016). Lapses in oversight of food supplies causing foodborne illnesses cost Americans more than $50 billion per year (Scharff, 2012). Thus, we cannot conclude that we are doing enough to protect people from unsafe food.

More than 250 foodborne diseases have been identified. Most of them are infections caused by a variety of bacteria, viruses, and parasites. In addition, harmful toxins and chemicals also can contaminate foods and cause foodborne illness. The most common foodborne illnesses are identified in Table 2.1. Four categories of microorganisms causing foodborne illnesses may be distinguished.

One group of microorganisms contains bacteria that include *Campylobacter, Salmonella, Listeria, Escherichia coli*, and *Yersinia*. Second, toxins from bacteria may lead to illness, including toxins of *Staphylococcus aureus, Clostridium perfringens, Clostridium botulinum*, and *Bacillus cereus*.

The third category of foodborne illnesses is viruses. These include norovirus, rotavirus, hepatitis A virus, and hepatitis E virus. Infected humans may transmit a virus to food items and contaminate food supplies. The fourth category of microorganisms causing illnesses consists of parasites. These include *Trichinella, Toxoplasma, Cryptosporidium*, and *Giardia*.

Food may become contaminated at different stages in the food chain. It may start at a farm when bacteria such as *Salmonella* find their way into food products. During the slaughter of food animals, meat can become contaminated with feces or pathogens, leading to diarrhea and gastrointestinal illnesses. Poor food preparation practices, such as contaminated contact surfaces or infected humans

Table 2.1 Common foodborne illnesses*

Germ	Organism	Common sources	Costs
Campylobacter	Bacteria	Raw or undercooked poultry	$1.9 billion
Clostridium perfringens	Bacterial toxin	Raw meat and poultry	$0.34 billion
Escherichia coli	Bacteria	Leafy vegetables	$0.27 billion
Listeria	Bacteria	Food products	$2.8 billion
Norovirus	Virus	Infected people, surfaces, contaminated raw products	$2.2 billion
Salmonella	Bacteria	Uncooked food products	$3.7 billion
Toxoplasma gondii	Parasite	Raw and undercooked pork and lamb	$3.3 billion
Vibrio vulnificus	Bacteria	Raw or undercooked seafood	$0.28 billion

* CDC, 2017; USDA, 2017, 2018a

handling food, can lead to contaminated products. The failure to fully cook meats or inadequate refrigeration may create conditions that allow harmful microorganisms to multiply and cause illnesses.

Addressing food safety

There are many aspects to keeping our food products safe. After the terrorist attack of September 11, 2001, one of the concerns was acts of bioterrorism in our food supply. Congress responded by passing a bioterrorism act in 2002. This Act requires domestic and foreign facilities that manufacture, process, produce, or hold food for human consumption to register with the Food and Drug Administration (FDA).

Given the large number of cases of food-borne illnesses, we rely on governmental actions against probable safety lapses. Products causing foodborne illnesses are adulterated, and firms selling these products are violating federal law. Both the US Department of Agriculture (USDA) and the FDA have authority to take action against adulterated products.

The FDA and USDA have regulations requiring establishments to notify the agency within 24 hours of learning or determining a product is adulterated or misbranded. However, for products regulated by the FDA, only reportable food issues that rise to the level of a Class I recall must be reported to the FDA's Reportable Food Registry. This means that many violations do not have to be reported. For meat and poultry products regulated by the USDA, the notification requirements are applicable to all adulterated or misbranded products regardless of their classification.

Enforcement action against all violators is not possible. Rather, the government has discretion over initiating enforcement proceedings under federal law. Responses include warning letters, import alerts, recalls, civil actions to impose money penalties, injunctions, and seizures of adulterated products. For noncompliance actions, the FDA and the USDA need to coordinate their efforts with the Department of Justice.

Perhaps the most drastic action against an offending article of food is seizure by the FDA. To affect a seizure, the agency needs to use a judicial process, and files a complaint against the food item requesting it be condemned and destroyed. A governmental seizure is rare, but enables action to prevent damages that could arise from the seized item.

For violations that present significant consumer safety issues, the government can take action to recall adulterated products. These may be voluntary or mandatory, but in most cases they are voluntary. Recalls serve to protect public health by removing adulterated and hazardous foods from commerce and ensuring fair trade by removing misbranded foods.

Food safety recalls are differentiated into three classes. Class I recalls involve a hazard with a reasonable probability that the product will cause adverse health consequences or death. An analysis of recent food product recalls shows that they mainly concerned the presence of undeclared allergens and the misbranding of products. However, the second leading cause is the presence of harmful pathogens. Class II and III recalls are less serious and involve potential health-hazard situations and situations that will not cause adverse health consequences.

An examination of meat and poultry recalls suggests they have not appreciably increased over the past two decades (Gorton and Stasiewicz, 2017). For the 131 recall cases handled by the USDA in 2017 concerning meat, poultry, and egg products, only 24 involved Class I recalls, involving *Escherichia coli*, *Listeria*, and *Salmonella* (USDA, 2018b). However, 53 cases involved undeclared allergens. Many of these recalls were also Class I recalls due to the serious adverse health consequences they created for persons who experience severe allergic reactions.

Although recalls have rather small effects on future demand for the products (Shang and Tonsor, 2017), they are costly for firms. A study of Class I meat recalls found that they negatively impacted the market value of the firms by 1.15 percent within five days following the recall (Pozo and Schroeder, 2016). Large recalls may lead to a firm's bankruptcy. To avoid costs, firms can strive to reduce the size of recalls by conducting product testing more frequently and on smaller lots.

Regulating food safety

The oversight of the safety of food products from animals is shared by the USDA and the FDA. The USDA is in charge of administering the Poultry Products Inspection Act, the Federal Meat Inspection Act, and the Egg Products Inspection Act. The USDA has designated the Food Safety and Inspection Service as its division in charge of the safety of animal products.

The dividing line between foods and their regulatory agencies tends to be confusing (Table 2.2). While there is a clear demarcation for beef, pork, lamb, and poultry products that are regulated by the USDA, other products are regulated by

Table 2.2 Identifying which federal agency regulates various food products

Food product	Agency	Comments
Beef	USDA	8 quality grades
Pork	USDA	"Acceptable" and "utility" grades
Lamb	USDA	5 quality grades
Poultry	USDA	3 quality grades
Seafood and catfish	FDA and USDA	Catfish regulated by the USDA
Eggs and egg products	USDA and FDA	Differentiate shelled eggs and egg products
Milk and milk products	FDA	Grade "A" and non-grade "A"
Meat sandwiches	USDA and FDA	Depends on the amount of meat
Raw fruits and vegetables	USDA	Often regulated by states
Processed fruits and vegetables	FDA	Includes fresh-cut fruits and vegetables

the FDA, and some have more subtle divisions. The significance is that under the USDA's legislative commands, all products must be inspected. The FDA is more likely to only inspect products after a tip about a food safety violation.

For products regulated by the USDA, they must be handled and held in a sanitary manner from production to consumption. Meat and poultry establishments develop and employ sanitation or processing procedures customized to the nature and volume of their production. All livestock and birds must be slaughtered humanely and must be inspected. More than 7,600 inspection program personnel are employed and assigned to more than 6,000 federal slaughter and food processing establishments.

The USDA's Food Safety and Inspection Service employs a science-based framework to identify and prevent food safety risks. Its framework is known as the Pathogen Reduction/Hazard Analysis and Critical Control Point (PR/HACCP) system, and it prevents the introduction of pathogens in the products we consume. Hazard Analysis and Critical Control Point addresses food safety through the analysis and control of biological, chemical, and physical hazards from raw material production, procurement, and handling, to manufacturing, distribution, and consumption of the finished product.

All segments of the food industry employ the HACCP system (Table 2.3). It incorporates prerequisite programs as foundations for the development and implementation of successful plans. The seven principles of HACCP have been universally accepted by government agencies, trade associations, and the food industry around the world. The system starts with the identification of hazards and the identification of appropriate control measures. The subsequent principles involve steps to effectuate features to keep food safe.

For products under the authority of the FDA, the provisions of the Food Safety Modernization Act and its Food Safety Plans govern activities to prevent foodborne illnesses. About two-thirds of the FDA's food-related resources are devoted

Table 2.3 Principles of the Hazard Analysis and Critical Control Point system

Principle		Action
Principle 1	Conduct a hazard analysis	Identify hazards and control measures
Principle 2	Determine the critical control points	Review ingredients, activities, equipment, storage, and distribution
Principle 3	Establish critical limits	Match practices and outcomes with the law
Principle 4	Establish monitoring procedures	Determine frequency
Principle 5	Establish corrective actions	Fix problems in a timely manner and alter procedures
Principle 6	Establish verification procedures	Verify the plan and record checks
Principle 7	Establish record-keeping and documentation procedures	Keep a diary and update written records

to addressing foodborne illnesses caused by microbiological pathogens in food (Taylor and Sklamberg, 2016).

The Food Safety Modernization Act provided the FDA the ability to require comprehensive, prevention-based controls across the food supply to prevent or significantly minimize the likelihood of problems occurring. It starts with hazard identification and the development of preventive controls. These include process controls with values or limits required for safety and food allergen and sanitation controls. A plan needs to record the oversight and management of preventive controls.

Each facility must also implement a risk-based supply-chain program if the hazard analysis identifies a hazard that requires a preventive control and the control will be applied in the facility's supply chain. The hazard analysis also needs a written recall plan. A program's provisions enable the agency to ensure that imported foods meet US standards and are safe.

Food additives

Meat and poultry products may also be treated with food additives (Table 2.4). Additives used include antioxidants to retard rancidity, enzymes to soften tissues, citric acid to protect the color of meat cuts, flavor enhancers, and phosphates for flavor protection. Food additives are regulated by the FDA.

Food additives are substances added to or affecting the characteristics of food that are not generally recognized as safe. Direct additives are purposefully added to food to serve a specific function. Indirect additives become part of food in very small quantities as a result of growing, processing, or packaging. Food additives include adding color to foods other than pigments from natural sources.

Food additives require premarket approval by the FDA before foods containing them can be sold. Conversely, foods containing added substances that are

Table 2.4 Common food additives added to meat products

Ascorbic acid and sodium ascorbate	Prevents oxidation that causes color change and spoilage
Bromelin and ficin	Used as meat tenderizers
Butylated hydroxyanisole	Prevents oils in foods from oxidizing and becoming rancid
Gelatin	Used as a thickener and in some canned ham and jellied meat products
Monosodium glutamate	Enhances flavors
Phosphates	To maintain moisture in products to enhance juiciness and tenderness
Sodium erythorbate	To maintain the color of processed meats
Sodium nitrite	To cure meat products and give them a characteristic pink color
Tocopherol	Prevents fats in meat and poultry from becoming rancid

"generally recognized as safe" need no further approval. Persons and firms intending to use a food additive are able to file a petition with the FDA that details sufficient information to establish that the food additive is safe and accomplishes its intended use. Once approved, the additive must be used within the constraints of its established regulation.

Food additives must serve a useful function. Manufacturers cannot use additives if the additives significantly decrease the nutritional value of the food or the manufacturer can obtain the desired effect using economical, good manufacturing practices without additives.

Generally recognized as safe

The use of a substance "generally recognized as safe" (GRAS) is one for which there is common knowledge by the scientific community that there is reasonable certainty the substance is not harmful under the conditions of its intended use. The determination of a GRAS use can be shown through scientific procedures or by general safety. If a manufacturer decides an added substance is a GRAS use, the substance does not need to undergo premarket review prior to product sales.

In 2016, the FDA promulgated a rule known as the GRAS Rule that amended the FDA regulatory criteria for determining when a substance is generally recognized as safe. The GRAS Rule establishes a voluntary generally recognized as safe notification process and prescribes a seven-part format for scientific data and other information in an applicant's notification. Manufacturers who chose to file an application for a GRAS use employ scientific procedures applying generally available and accepted scientific data, information, or methods.

The FDA issued a guidance for industry that helps manufacturers met the requirements for a GRAS use determination. Once the FDA receives a request, it reviews the manufacturer's data and informs the manufacturer whether the agency questions the basis of the generally recognized as safe determination.

Because applications for GRAS uses are voluntary, it remains controversial because manufacturers can add substances to food products without governmental oversight. In 2017, several public interest groups challenged the GRAS Rule, alleging it allows potentially unsafe food additives to enter the food supply without FDA review (*Center for Food Safety vs. Price*, 2017). The lawsuit seeks to require the FDA to review the safety of substances used in or in contact with food rather than allowing manufacturers to self-determine that a substance is safe so can be added without regulatory oversight.

Food waste due to safety concerns

It is estimated that 40 percent of the food we produce is never eaten (NRDC, 2017). The food is lost, with some remaining in fields, some deposited in landfills, and some used as animal food. We should be concerned about this wasted food because of the costs. The yearly losses have been valued at $218 billion (NRDC, 2017).

First, much of the waste occurs at the farm. About 7 percent of planted fields go unharvested each year (Moore, 2017). Numerous reasons account for decisions not to harvest produce, including low prices, inability to hire workers to harvest, weather conditions that markedly reduce the quality of the products, and safety lapses causing potentially adulterated products. Food produced but not used involves incurring costs for inputs and unnecessary uses of natural resources. This includes water, fertilizer, pesticides, and energy used to grow produce that is never consumed.

A large amount of the waste at the farm is related to USDA grades and standards (Friedman, 2017). Grades often require products be free of cosmetic damage, such as cracking, mechanical damage, weather events, and insect pests. Whenever there are no buyers for products that do not meet a standard, they are wasted.

Second, food waste constitutes a misallocation of investments in marketing. Firms abandoning food products lose money. They have expended funds to buy and transport the food and may need to pay to discard it. To recover these costs, firms need to charge higher prices for food products they sell. Many of the costs of wasted food are paid by consumers through higher prices for other food products. Consumers also waste considerable amounts of food and expend income for items they do not use.

It is estimated that food waste is one of the greatest sources of waste placed in landfills (NRDC, 2013). Between the wasted resources in producing food being thrown away and the methane being emitted from landfills containing food waste, we are needlessly engaging in activities that damage the environment and contribute to climate change.

Some food waste occurs because manufacturers place labels on food products with "use by" or "sell by" dates. Consumers read the date labels and assume the food item is unsafe if the date has passed. This is not true for many products. The dates placed on food by firms have various meanings. Some manufacturers use a date to define the shelf life of a product according to changes in product quality over time. Others define shelf life to mean the sheer absence of any decline in product quality. Dates used for these products are connected to product quality rather than product safety.

Because of sell-by and use-by dates, an estimated 90 percent of Americans occasionally throw edible food away out of a mistaken concern for food safety (NRDC, 2013). Due to legislative enactments by nearly 40 state legislatures and other rules, food products with an expired date cannot be given to charitable organizations. State legislatures have confused food quality with food safety to unnecessarily deprive people of food items.

Food manufacturers have proposed voluntary guidelines to limit date labels to "best if used by" to describe product quality (Painter, 2017). This would help consumers realize that using the product after the date label is safe. For perishable products, a "use by" label could be used. If these labels are adopted, and if consumers can learn that the "best if used by" label is not a food safety warning, we could reduce the amounts of food being wasted.

Food security

Food security involves having enough food to eat. Often this begins with people having enough money to buy food. While Americans are aware that millions of people in the world lack resources to secure sufficient food, they are not aware that food insecurity occurs in the United States. An estimated 41 million Americans lack adequate nourishment and are food insecure.

In addressing food security, our initial response is to provide people access to more food. Charitable organizations have performed admirably in getting food to millions of hungry people. However, access alone may not be sufficient to end hunger. Due to other circumstances, persons with access to food may still be food insecure because they lack access to clean water and wholesome food. The presence of contaminants causes people to become ill and sometimes leads to death.

Furthermore, the food supplies must provide minimal nutrition for people to be food secure. The absence of key vitamins or minerals in a diet causes malnourishment and stunting. The lack of proper nutrition also may contribute to medical problems and higher rates of disease, and may impair the cognitive development of children. An estimated 805 million people in the world are chronically undernourished (FAO, 2014).

For many people, products from animals are important in providing all of the essential amino acids in one protein source that can be easily used by the body. The nourishment provided by amino acids enables our bodies to manufacture new tissues and repair old ones. While beans, legumes, leafy greens, and other vegetables contain the nine essential amino acids needed by our bodies; due to eating habits, some people may not consume adequate portions to obtain needed amounts. Proteins derived from fish, poultry, eggs, and meat are generally complete and thus provide the needed amino acids to keep us healthy.

Finally, people need stability in their sources of future food. Lapses in the stability of food supplies occur due to drought, pest outbreaks, and income losses. In developing nations, livestock are used to cultivate fields to grow crops and are able to eat plants that are not a food source for humans. They provide high-value protein for consumers and generate income so owners can purchase other food products. Livestock also provide a buffer against crop failures and natural disasters and so contribute to food access and stability.

Many Americans are oblivious to the food insecurity problems experienced by people with low incomes who lose their job or have a medical emergency. Due to our focus on food safety, we fail to realize that some poor people are unable to purchase nutritious food.

Whatever the reason, food insecurity is a drag on our economy. Hungry people cannot make as meaningful contributions to our country's work force. Due to concerns about having enough to eat and diminished cognitive abilities, children who are food insecure will not be able to perform well in school. We need to assist people who are food insecure to obtain sufficient food so they can work. We need to provide children nutritious food that facilitates normal development and growth. We are failing to achieve our potential as a nation if we decline to make sufficient nutritious food available to every citizen.

Foodwashing facts

1   Approximately 41 million Americans lack adequate nourishment.
2   Foodborne illnesses cost our country more than $50 billion per year.
3   Many substances in food were never tested or authorized by the FDA.
4   Each year Americans waste food products valued at $218 billion.

References

Berman, J., et al. 2013. Can the world afford to ignore biotechnology solutions that address food insecurity? *Plant Molecular Biology* 83, 5–19.

CDC (Centers for Disease Control and Prevention). 2017. *Foodborne Illnesses and Germs.* www.cdc.gov/foodsafety/foodborne-germs.html.

Center for Food Safety vs. Price. Case 1:17-cv-03833 (Southern District of New York, filed 22 May 2017).

FAO (Food and Agriculture Organization of the United Nations). 2014. *The State of Food Insecurity in the World.* Rome.

FDA (Food and Drug Administration). 2016. *Foodborne Illnesses: What You Need to Know.* www.fda.gov/food/resourcesforyou/consumers/ucm103263.htm.

Freidman, E. 2017. Towards 2030: Shortcomings and solutions in food loss and waste reduction policy. *Washington University Journal of Law & Policy* 55, 265–293.

Gorton, A., Stasiewicz, M.J. 2017. Twenty-two years of U.S. meat and poultry product recalls: Implications for food safety and food waste. *Journal of Food Protection* 80(4), 674–684.

Moore, R. 2017. Nasty weather and ugly produce. Climate change, agricultural adaptation, and food waste. *Natural Resources Journal* 57, 493–516.

NRDC (Natural Resources Defense Council). 2013. *The Dating Game: How Confusing Food Date Labels Lead to Food Waste in America.* Washington, DC: NRDC.

NRDC. 2017. *Wasted: How America Is Losing up to 40 Percent of Its Food from Farm to Fork to Landfill,* 2nd ed. Washington, DC: NRDC.

Painter, K.L. 2017. Food firms to simplify expiration date labels. *Star Tribune,* Minneapolis, MN. February 21, Metro Ed., News, p. 7A.

Pozo, V.F., Schroeder, T.C. 2016. Evaluating the costs of meat and poultry recalls to food firms using stock returns. *Food Policy* 59, 66–77.

Scharff, R.L. 2012. Economic burden from health losses due to foodborne illness in the United States. *Journal of Food Protection* 75(1), 123–131.

Shang, X., Tonsor, G.T. 2017. Food safety recall effects across meat products and regions. *Food Policy* 69, 145–153.

Taylor, M.R., Sklamberg, H.R. 2016. Internationalizing food safety: FDA's role in the global food system. *Harvard International Review* 37(3), 32–37.

USDA (US Department of Agriculture). 2017. Cost Estimates of Foodborne Illnesses. Cost of Foodborne Illness Estimates for *Escherichia coli* O157.

USDA. 2018a. *Salmonella* and *Toxoplasma gondii* Are the Costliest Foodborne Pathogens. *Agri Marketing,* March 30.

USDA. 2018b. *Summary of Recall Cases in Calendar Year 2017.* Food Safety and Inspection Service. www.fsis.usda.gov/wps/portal/fsis/topics/recalls-and-public-health-alerts/recall-summaries.

3 The meat industry

Key questions to consider

1 Do Americans eat more beef or chicken?
2 What are the controversies concerning the production of food animals?
3 Does the meat industry work with governments in providing quality meat products?
4 What marketing changes have contributed to lower-price meat products?

Many Americans enjoy eating meat products – beef, pork, poultry, and seafood – at a meal. Steaks, hamburgers, chicken fingers, and hot dogs are common fare for entertaining and social gatherings. It is not surprising that Americans have one of the highest consumption rates of meat products in the world: about 200 pounds per year. The consumption is supported by a robust meat industry that uses various practices and technologies to reduce production costs.

The industry is concerned about food safety, as it realizes contaminated products lead to reductions in consumption. With respect to healthy food, the industry has its own viewpoints, as it uses antibiotics, hormones, and feed additives in the production of meat products despite evidence suggesting these inputs are not needed and may negatively affect human health.

These issues are covered in later chapters. However, a discussion of some general conditions, practices, and concerns about meat production can provide a foundation for looking at more specific issues. We also can observe how the industry has implemented technologies and altered production and marketing practices to maintain the central position of meat products in the diets of many Americans.

Consumption changes disclose that Americans have shifted to eating more poultry products than beef. Although steaks and hamburgers may be popular American foods, and hamburger fast food restaurants account for 30 percent of fast food industry sales, Americans have reduced their consumption of beef by more than 30 pounds in the past 40 years (USDA, 2017). Furthermore, in response to concerns about health and climate change, many Americans are reducing their meat consumption. Our consumption of meat products per person has decreased from its peak in 2006 (Table 3.1).

Table 3.1 Pounds of meat consumption per person by animal species in the United States*

Year	Beef	Pork	Broilers/ chicken	Turkey	Total red meat	Total poultry	Total meat and poultry
1965	74.7	51.5	36.4	7.6	133.9	44.0	177.9
1975	88.2	42.9	38.7	8.3	136.3	47.0	183.3
1985	79.0	51.5	52.5	11.6	133.8	64.1	197.9
1995	66.4	51.5	68.9	17.6	120.0	86.5	206.5
2005	65.4	49.6	86.4	16.7	116.5	103.1	219.6
2015	53.8	49.2	89.3	15.9	104.2	105.2	209.4

* National Chicken Council, 2018

Yet because of its high protein content, meat can help people cut calories from their diets. Moreover, many social settings continue to view meat as an important feature of a meal. This includes grilling, fast food (hamburgers, chicken fingers, and hot dogs), barbeques, Thanksgiving turkeys, and Christmas hams. But, as public opinion shifts to a desire to do more to curtail climate change, as has occurred in Western Europe, there could be further reductions in the consumption of meat products, especially beef. For health reasons, other consumers may alter their diets to consume more plant-based food products, which will also reduce meat consumption.

Environmental concerns

While many of the environmental concerns are covered in subsequent chapters, one that should be highlighted is the concern by some consumers about meat's carbon footprint. Animals raised for food are responsible for significant releases of gasses that contribute to climate change. Figure 3.1 shows the amounts of carbon dioxide released during the production of beef, pork, dairy, poultry, and egg products.

Beef, lamb, pork, chicken, and seafood account for 44 percent of our food footprint. Furthermore, the footprint of beef and lamb is double the amount from other meat products. These observations are leading some consumers to advocate reductions in meat consumption.

Production practices

In selecting meat products, some consumers are concerned about the practices used to produce the animals providing the products. Subsequent chapters evaluate practices concerning the production of meat products, humane treatment of animals, and giving animals space for moving around. These are followed by six chapters on animal production inputs that look at practices employing the use of antibiotics, hormones, feed additives, pesticides, breeding and cloning, and genetic engineering.

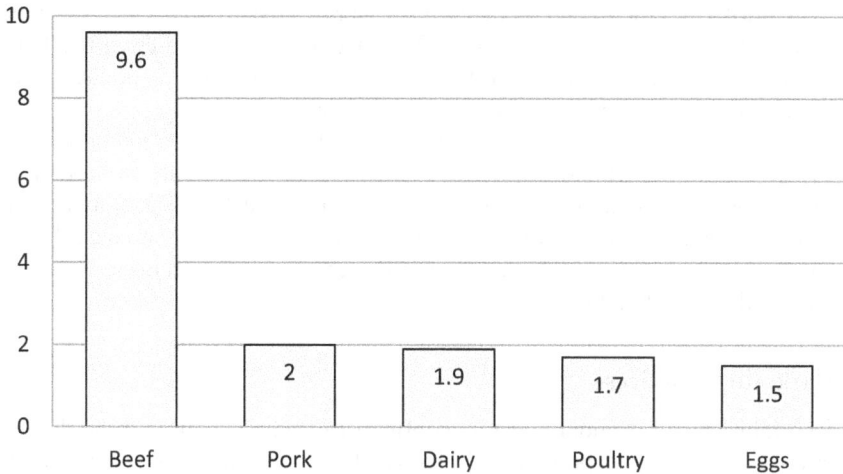

Figure 3.1 Equivalent carbon dioxide production related to 1,000 calories of animal products*

* Kunzig, 2014

Production practices affect the quality and taste of meat products. Consumers realize this fact and therefore often seek information about these practices, causing livestock producers to alter their practices to meet the concerns voiced by consumers. Selective breeding programs and the use of feed additives have led to products with less fat. Price premiums for quality products have led more humane production and handling practices. Other practices contribute to more tender meat products.

A few issues may be highlighted to show significant changes in production practices. First, American values have changed about the acceptable treatment of food animals. We are adopting European ideas on welfare standards and insisting that animals be treated humanely. This includes having sufficient space for animals to move during their confinement. Some producers are voluntarily engaging in additional practices that add production costs – such as free range – in order to garner higher prices for their products. Practices that were acceptable to our grandparents – such as methods of slaughter – may now be illegal.

Consumers have also been successful in having legislative bodies enact legislation addressing selected practices. Greater oversight and inspection of production facilities have led to healthier animals. Regulatory requirements on the use of antibiotics, hormones, and food additives limit harmful residues in food products. Various food safety and inspection requirements reduce the risks of contaminated meat products.

Another change in production involves the concentration of animals at large animal feeding operations. These aggregations have led many producers to adopt the use of antibiotics to augment animal growth and control diseases. The over-use of antibiotics for animal production has alarmed health officials as it may be contributing to antibiotic resistance. Other legislation has addressed the large amounts of manure containing nitrogen and phosphorus that is polluting our streams and water bodies to reduce water pollution.

Controls curtailing movement of food animals have become an important feature of controlling animal diseases. Veterinarians need to certify livestock and poultry for intrastate and interstate transportation. Health officials of individual states provide certificates of veterinary inspection that certify that animals do not carry illnesses or parasites. Vaccination requirements and quarantine also help prevent the spread of diseases.

Marketing practices

Technologies and economic factors continue to lead to changes in marketing practices. One of the pronounced changes has been the vertical integration of producers and marketers in the poultry and pork industries. Under vertical integration in the poultry industry, an integrator often owns the breeder flocks, hatcheries, feed mill, and processing plants. Integrators enter contracts with producer–growers, under which the integrator provides the chicks, feed, medication, and technical advice. Controls of feed inputs, genetics, and other important inputs in animal production have lowered production costs.

Integration means that producers have few choices in the production of their animals. Although they provide the land, labor, housing, equipment, and utilities, they generally do not have choices of when to receive new animals and when the animals will be scheduled for slaughter. Under integrated facilities, animals tend to be uniform, are produced in large numbers, and scheduling provides a steady supply of animals to the market. The sales prices are part of the contractual agreement, with a bonus for superior performance or a lower price for below-average performance. Integration is also accompanied by clustering of production facilities around processing facilities.

Following integration has been the consolidation of meat packing (slaughtering) facilities. Due to the lower prices for chicken products, the beef and pork industries analyzed their operations to find ways to cut costs. One way was to reduce expenses through economies of size. Large packers had advantages of greater capacity, more shifts, and lower costs in processing animals. More stringent food safety requirements entailed additional costs for packers that also favored large facilities. These cost advantages caused many small packers to go out of business.

Another change contributing to consolidation was the switch to boxed technology (Wohlgenant, 2013). Rather than shipping carcasses to wholesalers and retailers for cutting into final products sold to consumers, the cutting was done at processing plants. Because labor costs are lower at processing facilities than retail facilities, this allowed the industry to reduce the costs of retail products. Furthermore, when larger, nonunionized packing facilities opened in the Southern

Great Plains, many small packers in states with labor unions could not compete and closed.

Concerns exist whether the consolidation of packers means many producers have only a few potential buyers for their animals. Tyson Foods, Cargill Meat Solutions Corp., JBS USA, and National Beef Packing Co. control about 85 percent of beef processing. Four companies control about 60 percent of pork processing, and for poultry four companies control about 50 percent (USDA, 2016). Because there are few processing firms for producers to sell their animals, firms tend to have a monopoly and set the prices. Some analyses of prices suggest that there is a significant lack of marketing opportunities for producers in the Texas Panhandle regional cattle market.

The conditions at American meat processing facilities are dangerous. The Bureau of Labor Statistics reports injury and illness rates for the meat packing industry are double the national average (US Department of Labor, 2018). The US Government Accountability Office found that the meat and poultry industry had the eighth highest number of severe injury reports of all industries (GAO, 2017). Workers in meat and poultry slaughter and processing plants continue to face hazardous conditions, as inspections suggest some injuries are not being reported. An analysis of facilities found that vulnerable workers fear for their livelihoods and are pressured not to report injuries (GAO, 2016, 2017). Another observation was that the lack of coordination between worker safety inspection and food safety inspection regulations facilitates changes beneficial for food safety but detrimental for worker safety (GAO, 2017).

One technology that is still being developed is the tracing of animals. Tags for individual animals, mainly cattle and pigs, are available to identify the locations of their production and slaughter. Some cattle producers are using the tracking system to offer security against theft. Another use for tagged animals is to analyze anomalies in livestock behavior to identify diseased animals. Further use of blockchain technologies are advocated to supply product information to consumers.

Meat exports

Our producers of food animals are dependent on meat exports. Due to the use of modern production and processing practices, American meat products are competitive in international markets. Beef exports in 2017 were valued at $7.27 billion, pork exports were valued at $6.49 billion (US Meat Export Federation, 2018), and poultry exports were valued at nearly $4.6 billion (National Chicken Council, 2017). In 2017, approximately 12.9 percent of beef and 26.6 percent of pork produced in the United States were exported (US Meat Export Federation, 2018). The poultry industry exported about 18 percent of total production.

Exports of products from hogs include feet, elbows, and internal organs. These parts of a pig are not desired by American consumers, so the export market is important in securing greater income from marketing hogs. Exports of poultry products consist mainly of leg quarters and other back-of-bird parts that are not big sellers in the American market. By exporting these products, the industry is able to enhance its profits.

The export figures show that international trade is very important for related industries. The animals supplying exports consume more than 20 billion pounds of feed grains (US Meat Export Federation, 2018). Facilities and production inputs are needed for the animals providing these exports. Transport, slaughter, packing, and marketing facilities for meat exports are an important component of the American economy.

Important institutions

The production of our meat products involves considerable governmental oversight and assistance. While many acknowledge the safety requirements imposed by governmental laws and regulations, sometimes it is not realized that the USDA's National Institute of Food and Agriculture supports universities and local offices of the Cooperative Extension Service to provide research-based information to producers, businesses, and consumers. More than 100 land-grant colleges and universities are involved in this effort, and thousands of extension offices exist in counties across the country. The USDA also provides funding for animal health and disease research.

The USDA's Animal and Plant Health Inspection Service protects animals providing our meat products from pests and diseases. Under this service, the National Veterinary Services Laboratories help diagnose diseases and supports disease control and eradication programs. A National Veterinary Accreditation Program accredits veterinarians for certifying the health status of livestock, and they issue international health certificates needed for the export of animals to other countries.

Various public interest groups represent the meat industry, including dairy, poultry, and eggs. They have a number of purposes, but perhaps the most noteworthy is their lobbying activity. The industry spends more than $11 million a year lobbying on bills affecting their products (Sultan, 2017). Disease outbreaks such as avian influenza and mad cow disease can lead to temporary spikes in lobbying activities. Yet, the meat industry's outlays for lobbying are less than half being spent by food processing and sales, and about one-third of the amount spent by feed suppliers and agricultural pesticide companies.

In 2016, the industry took issue with the USDA's proposed new dietary guidelines. The advisory committee recommended that Americans should be encouraged and guided to reduce consumption of red and processed meat (USDA, 2015a). However, after intensive lobbying by the meat industry, the final guidelines failed to acknowledge limiting red meat products as part of a healthy eating pattern (USDA, 2015b). The USDA rejected science and the advice of its advisory panel. The dietary guidelines include lean meat as part of a healthy eating pattern and rejected consideration of environmental sustainability.

Health issues

Considerable concern exists whether persons eating meat are more likely to develop cancer or other diseases. In 2007, respected authorities published a report reporting that red and processed meats cause cancer of the esophagus and colorectal cancer (World Cancer Research Fund/ American Institute for Cancer Research, 2007). The International Agency for Research on Cancer, the cancer agency of the World Health Organization, classified processed meat as carcinogenic to humans and red meat as probably carcinogenic to humans.

Numerous other research has tied red and processed meats to prostate and other types of cancer. Red and processed meats are also related to type two diabetes and cardiovascular disease. A study involving more than 90,000 women found that higher consumption of red meat was associated with breast cancer incidences (Cho et al., 2006). This has led to advisories that people should lower their consumption of red and processed meat products.

Other research shows that supplementary hormones used in beef production, such as zeranol, may be related to cancer. Due to concerns about cancer, the European Union banned the use of hormones in animal production in 1981 and continues to exclude meat products from animals injected with hormones.

Yet, it may be argued that voluntary limitations on red meat intake are not advancing improved health outcomes. Meat often supplies micronutrients needed by humans, so the reduction of meat consumption may be accompanied by nutrient deficiencies. Some feel that our overzealous focus on limiting red meat consumption has distracted from effective nutrition strategies to address diseases associated with obesity and aging populations (Binnie et al., 2014).

Another health issue is fat. Some meat products contain considerable amounts of saturated fat that may be unhealthy. Meat consumption may increase the risks of type two diabetes, coronary heart disease, and strokes (Feskens et al., 2013; Wolk, 2017).

Quality beef products

Some consumers seek to purchase tender steaks and other specialized beef products. Quality is particularly important for beef products, so regulations recognize different grades, which sell at different prices. Producers strive to obtain the highest quality meats possible in order to maximize their income. Various practices, however, detract from quality and result in producers not achieving their full income potential.

The first issue is color. While meat color is not an important eating characteristic, it is a major factor in consumers' willingness to pay more for quality cuts. Most consumers prefer bright, cherry-red meat, so darker meat cuts do not command the highest prices (Killinger et al., 2004). Consumers also prefer the fat in their beef cuts be white rather than yellow, so beef from feedlots is preferred.

Dark colored meat is known in the industry as "dark cutting beef." Consumers prefer red beef, so beef cuts with a dark color command a lower price. Cattle under stress immediately prior to slaughter use their glycogen reserves, leading the pH of their meat to be higher than normal. The change in pH leads to darker colored meat. This occurs in 1 to 2 percent of beef carcasses. Dark cutting beef is often used in the food service industry where the raw products are cooked before reaching the consumer.

To avoid dark cutting beef, producers and transporters seek to reduce the stress experienced by cattle prior to slaughter. Conditions in transporting animals to the slaughterhouse are most important in moderating stress (Schwartzkoph-Genswein and Grandin, 2014). This includes avoiding sudden load noises, situations causing fear, overcrowding, food and water deprivation, and extreme changes in climatic conditions.

The bruising of cattle prior to slaughter also leads to dark cutting beef and inferior meat products. Bruised cattle cost the industry millions of dollars each year. Producers and transporters attempt to maintain conditions that minimize the risks of bruised meat cuts. Producers raise animals without horns, castrate male animals so they are less aggressive, prevent crowding of animals, and refrain from rough handling practices.

The beef industry was presented with a controversy concerning the sale of lean finely textured beef, which some referred to as "pink slime." The product is processed beef renderings that are derived from trimmings that would otherwise be used for some inferior purpose or discarded (Adams, 2014). The USDA approved the product in 2001 for human consumption. However, when an ABC News series was aired in March 2012 referring to lean finely textured beef as "pink slime," the public reacted negatively. Sales of lean finely textured beef products plummeted.

In fact, the manufacturer of lean finely textured beef (Beef Products Inc.) lost 80 percent of its business within a month and decided to close three-fourths of its processing facilities (Reid, 2014). Beef Products Inc. sued the American Broadcasting Companies for statutory and common law product disparagement, defamation, and tortious interference. The lawsuit was settled in 2017 for $177 million. While the industry and government attest that finely textured beef is safe for human consumption, this is not the critical issue involving sales of this product. Consumers with reservations about pink slime prefer other meat products (Dupre et al., 2018).

Cultured meat and plant-based substitutes

Two developments gaining momentum are cultured meat and plant-based meat substitutes. Cultured meat is also known as clean meat, simulated meat, *in vitro* meat, or meat grown in a lab. It is an emerging technology that can provide high quality protein with a relatively small ecological footprint (Bekker et al., 2017). These products are isolated from cell cultures or tissue cultures derived from livestock and poultry animals or their parts.

Persons producing cultured meat claim it is healthier than conventional meat and more environmentally friendly. Health benefits may accrue from the absence of heme iron that is often in meat products. Heme iron is related to DNA damage and an increased risk of breast cancer. Yet, there is much to do to convince consumers that cultured meat products provide the nutrients and taste of real meat products.

The environmental benefits of cultured meat are appealing. It reduces the land and water needed to produce food and the numbers of animals needed to supply meat products, with a corresponding reduction in the release of greenhouse gasses. Cultured meat also reduces the use of antibiotics and hormones used in animal production. The food preferences of millennials show a decrease in meat consumption. Cultured meat products will displace meat in some products and at some meals.

Yet, cultured meat faces a number of obstacles. Perhaps the biggest challenge is that consumers are generally not fond of products that are not natural (Siegrist et al., 2018). Can cultured meat be marked as a natural meat substitute despite its various constituents? For some products, including burgers, flavor is the issue. People may find the taste of cultured meat to be inferior to real meat. Yet opportunities exist. For example, a cultured product that tastes like foie gras can enable consumers to have a tasty product that is not related to the inhumane treatment of ducks and geese.

The second substitute for meat products is to use plant products to create meat alternatives. The firm Beyond Meat has introduced a number of products to replace meats including "The Beyond Burger®," "Beyond Sausage™," "Beast Burgers," "Beyond Chicken® Strips," and "Beyond Beef® Crumble" (Beyond Meat, 2018). In 2018, an entrepreneur in Hong Kong announced the development of Omnipork, a plant-based product that can replace pork (Lamb, 2018). The product can be steamed, fried, or made into patties and could help China reduce its pork consumption.

The animal industry is concerned about cultured meats and substitutes and is advocating that the federal government take action with respect to the labeling of lab-grown imitation products. The industry is seeking new regulations to preclude cultured meat from being labeled as "beef" or "meat." In 2018, a petition was submitted to the Food Safety and Inspection Service with two major requests concerning the labeling of meat products (US Cattlemen's Association, 2018). The association requested that the government define "beef" to mean products coming from cattle that have been born, raised, and harvested in the traditional manner. A second request was that "meat" be defined to products that are the tissue or flesh of animals that have been harvested in the traditional manner. Both of these definitions would exclude products grown in labs from animal cells and from plants. Meanwhile, a state legislature may attempt to define what products can be labeled as meat (Missouri Senate Bills, 2018).

Foodwashing facts

1 Many consumers consider the consumption of meat products to be an important part of their diets and lifestyles.

2   Production practices for food animals have changed in response to consumer preferences.
3   Health issues have been related to the consumption of red and processed meat.
4   Consumers concerned about climate change and desiring to eat more plant-based food may curtail meat consumption.

References

Adams, R.J. 2014. Consumer deception or unwarranted product disparagement? The case of lean, finely textured beef. *Business and Society Review* 119(2), 221–246.

Bekker, G.A., et al. 2017. Explicit and implicit attitude toward an emerging food technology: The case of cultured meat. *Appetite* 108, 245–254.

Beyond Meat. 2018. *100% Plant-Protein: Our Products.* El Segundo, CA. http://beyondmeat.com/products.

Binnie, M.A., et al. 2014. Red meats: Time for a paradigm shift in dietary advice. *Meat Science* 98, 445–451.

Cho, E., et al. 2006. Red meat intake and risk of breast cancer among premenopausal women. *Archives of Internal Medicine* 166(20), 2253–2259.

Dupre, S.M., et al. 2018. Preference evaluation of ground beef by untrained subjects with three levels of finely textured beef. *PLOS One*, https://doi.org/10.1371/journal.pone.0190680

Feskens, E.J.M., et al. 2013. Meat consumption, diabetes, and its complications. *Current Diabetes Reports* 13, 298–306.

GAO (Government Accountability Office). 2016. *Workplace Safety and Health: Additional Data Needed to Address Continued Hazards in the Meat and Poultry Industry.* GAO-16-337.

GAO. 2017. *Workplace Safety and Health: Better Outreach, Collaboration, and Information Needed to Help Protect Workers at Meat and Poultry Plants.* GAO-18-12.

Killinger, K.M. 2004. Consumer visual preference and value for beef steaks differing in marbling level and color. *Journal of Animal Science* 82, 3288–3293.

Kunzig, R. 2014. Carnivore's dilemma. *National Geographic.* November. www.nationalgeographic.com/foodfeatures/meat/.

Lamb, C. 2018. Pig, out: Omnipork hopes to replace China's most consumed meat. *The Spoon.* https://thespoon.tech/pig-out-omnipork-hopes-to-replace-chinas-most-consumed-meat/.

Missouri Senate Bills. 2018. Nos. 627 & 925.

National Chicken Council. 2017. *NAFTA a Boon for US Poultry Exports, Say Top US Poultry Groups.* Washington, DC.

National Chicken Council. 2018. *Per Capita Consumption of Poultry and Livestock, 1965 to Estimated 2018, in Pounds.* Washington, DC. www.nationalchickencouncil.org/about-the-industry/statistics/per-capita-consumption-of-poultry-and-livestock-1965-to-estimated-2012-in-pounds/.

Reid, R.-M.C. 2014. You say 'Lean finely textured beef,' I say 'pink slime.' *Food and Drug Law Journal* 69, 625–647.

Schwartzkoph-Genswein, K., Grandin, T. 2014. Cattle transport by road. In *Livestock Handling and Transport*, Temple Grandin (ed.), 4th ed., pp. 143–173. Croydon, UK: CAB International.

Siegrist, M., et al. 2018. Perceived naturalness and evoked disgust influence acceptance of cultured meat. *Meat Science* 139, 213–219.

Sultan, N.M. 2017. *Where's the Beef? When Meat's in Trouble, Lobbying Expands.* Center for Responsive Politics. OpenSecrets.org.

US Cattlemen's Association. 2018. *Petition for the Imposition of Beef and Meat Labeling Requirements: To Exclude Products Not Derived Directly from Animals Raised and Slaughtered from the Definition of "Beef" and "Meat."* February 9. www.fsis.usda.gov/wps/portal/fsis/topics/regulations/petitions.

US Department of Labor. 2018. *Meatpacking: Overview. Occupational Safety and Health Administration.* https://www.osha.gov/SLTC/meatpacking/index.html

US Meat Export Federation. 2018. *FAQ.* www.usmef.org/about-usmef/faq/.

USDA (US Department of Agriculture). 2015a. *Scientific Report of the 2015 Dietary Guidelines Advisory Committee.*

USDA. 2015b. *Dietary Guideline for American 2015–2020.* 8th ed.

USDA. 2016. *Packers and Stockyards Program 2016 Annual Report.* Grain Inspection, Packers and Stockyards Administration.

USDA. 2017. *U.S. Per Capita Availability of Red Meat, Poultry, and Fish Lowest Since 1983.* Economic Research Service.

Wohlgenant, M.K. 2013. Competition in the US meatpacking industry. *Annual Review of Resource Economics* 5, 1–12.

Wolk, A. 2017. Potential health hazards of eating red meat. *Journal of Internal Medicine* 281, 106–122.

World Cancer Research Fund/American Institute for Cancer Research. 2007. *Food, Nutrition, Physical Activity, and the Prevention of Cancer: A Global Perspective.* Washington, DC: AICR.

Part II

Concerns about animal production facilities

4 Producing beef, dairy, and pork products

Key questions to consider

1 Why does the production of food animals involve the use of so many technologies?
2 What has helped reduce production costs?
3 Are animal health and well-being important to producers?
4 Are producers addressing the environmental issues presented by their production practices?

The production systems used to produce beef, dairy, and pork products have changed from the systems used on family farms of past generations. Images of beef cows foraging on the open range, dairy cows resting under shade trees, and pigs rutting in the soil are not accurate representations of how most of these animals are produced. Rather than having a number of species of animals, most farms raise one. The large numbers of animals allow producers to take advantage of technologies, economies of scale, and expertise.

Most beef products are harvested from animals that were fed in a feedlot. Although cattle may spend a large percentage of their lives on the range or in a pasture, they are moved to a feedlot for the last three to six months of their lives. During this period, they are fed high-caloric grains to help them gain weight. In many cases, they receive hormone implants and are administrated antibiotics. After they have gained enough weight, cattle are sent to slaughterhouses.

The production of milk products involves the use of technology to minimize human labor requirements. Cows are milked in milking parlors where a few people maintain the safe and efficient collection of milk. Milk is quickly transported to a processing facility and thereafter to market. Animal manure is removed as a solid or slurry and applied on fields as fertilizer for crop production.

Most hogs are produced in large barns. Sows are confined and carefully monitored during their farrowing period, when they give birth to 10–12 piglets. Pigs are moved to a nursery and later to a facility where they are raised as feeder pigs. Most pigs receive a specialized diet to maximize weight gain. Under good conditions, they are ready for harvesting in about six months.

Beef production

Cattle production for supplying consumers with beef products is one of the most important industries in the United States, accounting for about $78 billion in cash receipts per year. For a comparison, corn receipts often hover around $47 billion. Our country has about 90 million head of beef cattle, and in addition, culled dairy cows provide beef products. Beef production is important not only on the Great Plains but also in many other states. Three production systems exist to take advantage of resources and technologies: cow-calf, stocker, and feedlot-finishing operations.

Some cattle operators choose to raise purebred cattle. The most popular breed is Angus, followed by Charolais and Hereford. Purebred animals can be sold at special purebred sales and command higher than average prices. They also serve as sires or dams for crossbred cattle that often gain weight faster.

Cow-calf operations provide the calves that will subsequently be raised for market or used as breeding stock. About 80 percent of cow-calf operations have fewer than 50 animals, as many part-time operators engage in raising cattle. At most operations, the mother cow will nurture her calf for at least 90 days. The availability of milk from their mothers allows calves to thrive and gain weight quickly.

Many cow-calf operators select a 60–90-day period of time for breeding. Artificial insemination may be used, or a bull may service the heifers and cows. With a controlled breeding season, all of the calves are born around the same time. This facilitates attaching ear tags, weaning, vaccination, castration, and deworming programs to augment animal health. Three-fourths of bull calves are castrated before three months of age to control aggression and improve overall health and weight gain.

With lots of small producers, auction barns remain an important market for weaned calves. These calves are bought by producers to be raised as stocker cattle. Producers of stocker animals place the animals on pasture but may need to feed the animals hay when pastures cannot supply sufficient feed.

Prior to the transfer of cattle to a feedlot, animals receive vaccinations and boosters to control diseases. They also are preconditioned to reduce the likelihood that they become ill. This includes training the animals to eat grain from a bunk and drink water from an automatic waterer.

At a feedlot, cattle consume high-energy, grain-based diets. Modern facilities with more than 50,000 head have economies of scale so that the cost of feeding each animal is low. Feedlot operators usually use implants of growth stimulants to increase muscle growth and overall carcass weight.

Feedlots receive a lot of negative publicity due to their characteristics. They are smelly and generate a lot of manure that needs to be disposed without causing water pollution. The enclosure of many animals in a limited space prevents animals from expressing their normal behavior. In addition, some people object to the use of grain as they feel it should be used for human consumption and that cattle should be eating grass. However, feedlots lower the costs of producing meat products and provide products desired by a majority of the public.

Grass-fed beef

Some people feel that grass-fed beef should be preferred due to human health advantages, animal welfare, and environmental attributes. Grass-fed beef is currently a small percentage of the industry in the United States, making up about 4 percent of all beef produced. Since the USDA does not oversee labels identifying grass-fed meat, private certification firms are used so consumers know a product meets the labeling claim.

A reason that the grass-fed beef sector remains small is due to differences in appearance and taste of the meat products. Grass-fed beef has less fat due to the increased mobility of the animals and their lower caloric intake. Because the meat of these animals does not acquire as much marbling as grain-fed beef, it is not as tender or juicy. The fat that does exist in the beef is typically more yellow-colored than in its conventional counterpart, which some consumers find unappetizing (Leheska et al., 2008).

However, grass-fed beef is a healthier option due to having fewer calories and lower levels of saturated fat and total fat. It also is less likely that the animals have been produced with hormones, feed additives, or antibiotics. However, the grading system used by US Department of Agriculture (USDA) identifies beef with significant fat content and marbling as "prime." This grading system places a lower value on lean, grass-fed beef.

Some consumers believe that the production of grass-fed beef treats animals in a more ethical manner and is an environmentally friendly production system. Animals are able to roam in a pasture, so they have a higher quality of life than those confined in feedlots (Siegford et al., 2008). Grass-fed systems have fewer environmental problems associated with the disposal of large quantities of manure. Because corn is not fed to the cattle, few herbicides are used, and there is less soil erosion.

Yet, there are negative features connected to grass-fed animals. Animals that consume grass rather than grain products produce increased quantities of methane (Capper, 2012). Animals raised on forage diets also take longer to gain weight and have longer life spans before they are harvested as compared to feedlot-raised cattle.

Dairy cows and milk production

Recent USDA data estimate that our country has 9.39 million dairy cows at 40,219 farms producing more than 215 million pounds of milk each year (USDA, 2018a). While these data suggest that the average dairy farm has around 233 cows, large variations exist. For example, California, the leading dairy state, has 1,757,000 cows at 1,390 farms, meaning each farm averages 1,264 cows (USDA, 2018a).

Most dairy cows in the United States are Holsteins, a black and white breed known for producing more milk than other breeds. A Holstein can produce up to seven gallons of milk per day. To maintain production, a cow must eat about 100 pounds of food per day. Optimum milk production depends on a number of factors, including comfortable housing, plenty of water, and a nutritional feeding program. Cows are monitored to make sure they are healthy to maximize milk production.

Dairy cows are bred each year, almost always by artificial insemination, so that they continue giving milk. Approximately ten months after giving birth to a calf, a cow's milk output decreases substantially. Most cows enter a "drying off" period, during which they are not milked so they can rest. This practice optimizes subsequent milk production. Although cows have an expected life span of 20 years, most dairy animals are culled after four years and marketed for beef.

There is a trend toward larger and more efficient dairy farms. Smaller farms have lower milk yields per cow and higher average costs of production. The USDA's Economic Research Service estimated that the smallest farms with fewer than 50 cows had costs that were about double those of farms with 1,000 cows or more (USDA, 2018b). Statistics suggest that the more efficient farms are more likely to increase their herd sizes (Dong et al., 2016).

Milk production is labor-intensive because of the need of proper care for the cows, the need to prevent harmful organisms from adulterating the product, and its perishability. Comprehensive and strict regulations oversee the production of milk to ensure its wholesomeness. In addition, the concentration of animals at farms creates large amounts of manure and ammonia that may negatively affect the environment.

Housing for dairy animals

Dairy farms have choices when choosing housing for their cows. Housing that allows natural expression of animal behavior, maintains cleanliness, and ensures cow health can contribute to the economical production of milk. Cows need to be able to lie down, and some prefer to rest 14 hours a day. They must be kept clean during this time, so housing is structured so each cow can lie in an area free from manure and urine. Options for housing dairy animals include bedded-pack, tiestall, or freestall barns.

Bedded-pack barns involve an area for cows to rest that is bedded with sawdust or fine wood shavings. The bedding is composted in place with manure and maintained so it is a dry resting place for cows. New bedding is continuously added, which can be costly (Bewley et al., 2017). The bedding is typically stirred while the cows are being milked to facilitate drying. Cows feed, drink water, and are milked in separate areas, so that the manure and urine from feeding and milking areas are handled separately.

In tiestall housing, individual stanchions are used for each cow. Small operations with under 50 animals generally use this system. With the cows in stanchions, the milking machines are brought to the cows, with the milk collected via a pipeline (Douphrate et al., 2013). Workers milking the cows have to squat near each cow to attach the milking equipment. A gutter at the rear of the animals collects manure and urine, which can be removed. Individual automatic waterers are installed for one or two cows.

Freestall barns allow cows to rest in areas separate from where they eat and are milked. Cows go to a milking parlor to be milked and exit to an area to feed. Resting areas are designed so that animals are not lying in manure or wet bedding, and each cow is free to select a place for resting. For most freestall systems, large quantities of water are used for sanitation and flushing animal waste into lagoons. Nearly all herds with more than 200 dairy cows use freestall barns.

Technologies used at dairy farms

The producers of milk have found that the adoption of new technologies can help reduce costs. Choices of housing systems have already been discussed. Producers also use milking systems, individual electronic identification of cows, and sensors to evaluate animal health (Douphrate et al., 2013). With so many technological options, each producer has to make decisions about which technologies to use. Technologies involving vaccination, dehorning, nutritious grain mixes, and transition shelters are important for weaned calves. When heifers are 13–15 months old, they are artificially inseminated, so they can start producing milk by 24 months of age.

Cows are milked two or three times a day, and two milking parlor systems are common for bedded-pack and freestall dairies. The first parlor system is a traditional parlor at which milking machines are attached to cows by workers in a pit. By being lower than the cows, workers do not need to bend over to attach milking machines to the cows. A worker can operate multiple milking machines so that more than one cow is being milked at the same time. Typically, one person is employed for every 80–100 cows.

The second parlor system uses an automated computer-controlled system for attaching and detaching milking machines to the cows. Although these systems are expensive, they reduce labor costs and increase amounts of harvested milk from the herd. An automatic milking system allows cows choose when they want to be milked, and uses technology to do the milking. A cow enters the system, her teats are treated to remove impurities and promote milk let down, and the milking machine cups attach themselves to the cow to harvest the milk. After the milk has been harvested, the cups remove and clean themselves so they are ready for the next animal. A post-milking spray is applied to protect the udders against bacteria.

Dairy farmers use electronic identification (EID) tags. These are small button-like tags that are placed in the animal's ear and contain a unique number. Tags are designed to last for the life of the animal. An EID reader can scan and read a tag's number. With an EID system, a farmer can automatically collect a variety of data for each animal, including milk yield, milking time, activity level, weight, health problems, and reproductive status. The information enables individual animals to be sorted as they enter or leave milking parlors and feeding areas. In Canada, all dairy cows have a national livestock identification tag.

Another technology that offers potential benefits for animal producers is a wireless sensor network application (Kiani, 2018). Simple applications can distinguish three types of animal behavior: standing, walking–grazing, and lying down. Other systems can measure the feeding time of dairy cows and predict health issues such as lameness and bacterial infections (Pastell and Frondelius, 2018). The use of these technologies can help producers identify issues that detract from animal performance and to take appropriate action to remedy problems that adversely affect animal well-being.

Hog production

In the US, most hogs are raised in confinement barns that contain hundreds of animals. More than 120 million hogs are marketed each year (USDA, 2018c). Generally, three types of swine production systems are used: farrow-to-finish, farrow-to-feeder, and feeder-to-finish. The farrow-to-finish system means the piglets are born on the producer's farm and animals are raised until they are ready to go to market. The entire production period includes four months for breeding and gestation and six months for the animals to grow to marketable weight.

Sows are generally artificially inseminated, and gestating sows may be held in crates. However, most new operations use pens so sows have greater freedom of movement. Just before farrowing, sows are moved to a barn, where they are housed until the piglets are weaned. The sows are immediately bred again so they can have a second and third litter in a given year. Baby pigs are weaned at two to four weeks of age.

Weaned pigs are moved into a nursery, where they stay until they weigh 40–50 pounds. While at the nursery, the piglets are fed a corn and soybean meal diet. A piglet may eat as much as 4 pounds of feed a day. After they are sufficiently developed, they are moved to a finishing barn and fed until they reach marketable weight of 240–280 pounds. Feeder pigs eat 6–10 pounds of feed a day.

A farrow-to-feeder operation involves breeding and farrowing sows and raising the piglets in a nursery until they weigh 30–60 pounds. The piglets are then sold to a feeder-to-finish operation. The feeder-to-finish producer buys feeder pigs and raises them to market weight. In most cases, additional ingredients are added to feed to achieve a nutritious diet for the animals. Most feeder pigs are raised under a contract for a major pork company.

Pig behavior provides information about their health and welfare that affects production efficiency. One technology is a Respiratory Disease Monitor that monitors pig coughs (Matthews et al., 2017). The cough index is a measure based on the number of coughs across a group of pigs in a day. This technology can inform a producer that something is adversely affecting pigs, such as an uncomfortable building temperature or a ventilation problem.

As noted for dairy cows, automated monitoring of animal behavior may be used to detect animals resting, feeding, drinking, movement, and engaging in aggressive behavior (Nasirahmadi et al., 2017). Charge Coupled Devices (cameras) can monitor animal behavior, and the application of modern digital technologies in

3D imaging systems are able to predict health issues. Producers can use these data to quickly address problems and apply interventions.

Hog facilities and environmental problems

Economies of scale have led to hog production facilities with large numbers of animals. While these facilities facilitate the profitable production of pork products, they also are accompanied by characteristics that are controversial and problematic. The major issue is the smells that accompany hog production. With thousands of animals at one location, they generate a lot of waste. The facilities and anaerobic decomposition of swine waste result in airborne gasses that are objectionable. Hydrogen sulfide and ammonia are two common chemicals produced by hog waste, but hundreds of other chemicals are present in the odors from hog farms.

Production facilities usually consist of large buildings built over a pit that collects the waste. Slotted concrete floors allow the urine and manure from the pigs to drop or be washed into the pit. Pits are flushed one a week into a holding tank or lagoon. The contents can later be applied to cropland as a fertilizer.

The barns, lagoons, and fields emit odors into the air. With a wind, these are quickly carried to neighboring properties and in some cases are highly offensive. The construction of new facilities can be unfair to existing residents in an area. Due to the terrible smells, neighbors may sue producers for a nuisance and request relief from the smells. However, state anti-nuisance provisions (called "right-to-farm" laws) often prevent any remedy.

Yet, state anti-nuisance laws were not intended to allow new facilities to impose smells on neighbors. Recently, a North Carolina court recognized that the state's anti-nuisance defense does not apply in situations where no changed conditions are causing a hog farm to become a nuisance. Neighbors may also advance arguments that operators are not using best management practices, are using outdated technologies, have designed their waste facilities incorrectly, or lack appropriate equipment to manage wastes effectively.

Efforts are underway to ameliorate the odors coming from hog facilities. Through the use of anaerobic digesters, odors can be reduced (Adair et al., 2016). Other ideas are also being advanced to control pathogens, odors, and emissions of ammonia and greenhouse gasses from hog facilities (Adair et al., 2016). The traditional lagoons used to collect waste from hog operations may involve outdated technology. Environmentally superior technologies need to be installed at hog facilities to reduce their objectionableness.

Feral swine

In many rural areas, wild pigs known as feral swine thrive and create problems for agricultural producers and property owners. Numbers of feral swine can increase quickly, as these animals are adaptable to their surroundings. An estimated five or six million feral swine inhabit 1,323 counties in 39 states (Holderieath et al., 2018).

Because feral swine carry diseases and damage agricultural crops, they are unwelcome. They serve as reservoirs for pathogens which might be transmitted to domestic pigs (Hill et al., 2014). As a source of influenza A viruses, feral swine potentially could transmit the virus to domestic swine or humans (Feng et al., 2014). Feral swine are also detrimental to corn and peanut crops in Texas and the Southeast. It is estimated that annual economic crop losses due to feral swine in 11 states are $190 million (Anderson et al., 2016).

Feral swine can have a significant impact on native species and ecological resources. Rooting by feral swine harms grasses foraged by cattle and alters plant communities. Their activities reduce plant species and thus pose an ecological and economic threat to existing agricultural activities. Feral swine eat invertebrates, salamanders, frogs, snakes, turtles, fish, crabs, rodents, and muskrats. They also adversely impact ground-nesting birds and seriously threaten the nest success of several threatened and endangered sea turtles.

Meat product traceability

Given the public concern about safe food, recommendations have been made to develop a traceability system that would quickly identify tainted meat products so they could be removed from the market. Cattle producers already use electronic identification tags for their animals. This system could be extended so that meat products could be traced in the marketplace.

Data-capture technologies, such as radio frequency identification and two-dimensional barcodes, provide technologies that could be used to develop a dynamic traceability system (Buskirk et al., 2013). Alternatively, a combination of specialized technologies could be used to trace meat products (Wu et al., 2018). These include blockchain technologies so that meat can be traced through the supply chain. This provides marketers information for reducing risks and improving food safety. Consumers might garner more information on production practices.

The reluctance of the meat industry to adopt traceability involves privacy, competitiveness, and cost. Producers worry that traceability technology would enable contaminated products to be traced to their farms so that they could become liable for damages. While producers may enjoy a certain degree of anonymity, experiences involving tainted fruits and vegetables show that this may not be true. Producers and marketers need to accept responsibility for their practices and lapses in safety.

Moreover, our top priority should be food safety. By declining to do more in facilitating the traceability of meat products, we are limiting our ability to quickly identify the source of a problem (Mitchell, 2018). A corresponding difficulty could be the loss of competitiveness in the world market. As other countries adopt meat traceability safety requirements, we may not be able to meet the expectations of their consumers. Since we export more than one-quarter of our pork and about 13 percent of our beef, we ought to be adopting new technologies known to enhance food safety.

Foodwashing facts

1   Producer efficiencies enable American farmers to produce ample supplies of low-cost meat products.
2   Producers are using technologies to maintain the health and well-being of animals.
3   Additional efforts may be needed to address issues related to the disposal of animal waste.
4   The American meat industry is gradually enhancing the safety of meat products through the adoption of additional traceability technologies.

References

Adair, C.W., et al. 2016. Design and assessment of an innovative swine waste to renewable energy system. *Transactions of the American Society of Agricultural and Biological Engineers* 59(5), 1009–1018.

Anderson, A., et al. 2016. Economic estimates of feral swine damage and control in 11 US states. *Crop Protection* 89, 89–94.

Bewley, J.M., et al. 2017. A 100-year review: Lactating dairy cattle housing management. *Journal of Dairy Science* 100, 10418–10431.

Buskirk, D.D., et al. 2013. A traceability model for beef product origin within a local institutional value chain. *Journal of Agriculture, Food Systems, and Community Development* 3(2), 33–43.

Capper, J.L. 2012. Is the grass always greener? Comparing the environmental impact of conventional, natural and grass-fed beef production systems. *Animals* 2, 127–143.

Dong, F., et al. 2016. Technical efficiency, herd size, and exit intentions in U.S. dairy farms. *Agricultural Economics* 47, 533–545.

Douphrate, D.I., et al. 2013. The dairy industry: A brief description of production practices, trends, and farm characteristics around the world. *Journal of Agromedicine* 18, 187–197.

Feng, Z., et al. 2014. Influenza A subtype H3 viruses in feral swine, United States, 2011–2012. *Emerging Infectious Diseases* 20(5), 843–846.

Hill, D.E. 2014. Surveillance of feral swine for Trichinella spp. and Toxoplasma gondii in the USA and host-related factors associated with infection. *Veterinary Parasitology* 205, 653–665.

Holderieath, J.J., et al. 2018. Valuing the absence of feral swine in the United States: A partial equilibrium approach. *Crop Protection* 112, 63–66.

Kiani, F. 2018. Animal behavior management by energy-efficient wireless sensor networks. *Computers and Electronics in Agriculture* 151, 478–484.

Leheska, J.M., et al. 2008. Effects of conventional and grass-feeding systems on the nutrient composition of beef. *Journal of Animal Science* 86, 3575–3585.

Matthews, S.G., et al. 2017. Early detection of health and welfare compromises through automated detection of behavioural changes in pigs. *The Veterinary Journal* 217, 43–51.

Mitchell, R. 2018. Without traceability, U.S. beef could be left out in the cold. *The National Provisioner*, February, pp. 24–26. www.provisioneronline.com/articles/105844-without-traceability-us-beef-could-be-left-out-in-the-cold.

Nasirahmadi, A., et al. 2017. Implementation of machine vision for detecting behaviour of cattle and pigs. *Livestock Science* 202, 25–38.

Pastell, M., Frondelius, L. 2018. A hidden Markov model to estimate the time dairy cows spend in feeder based on indoor positioning data. *Computers and Electronics in Agriculture* 152, 182–185.

Siegford, J.M., et al. 2008. Environmental aspects of ethical animal production. *Poultry Science* 87, 380–386.

USDA (U.S. Department of Agriculture). 2018a. *Milk Production. National Agricultural Statistics Service.* ISSN: 1949–1557.

USDA. 2018b. *Milk Cost of Production Estimates. Economic Research Service.* www.ers.usda. gov/data-products/milk-cost-of-production-estimates.aspx.

USDA. 2018c. Livestock and poultry: World markets and trade. *Foreign Agricultural Service.* https://apps.fas.usda.gov/psdonline/circulars/livestock_poultry.pdf.

Wu, Q., et al. 2018. Multiplex TaqMan locked nucleic acid real-time PCR for the differential identification of various meat and meat products. *Meat Science* 137, 41–46.

5   The production of chickens

Key questions to consider

1   Have consumers been successful in getting more space for laying hens?
2   Are free-range hens healthier than other hens?
3   Are eggs from cage-free hens more likely to be free from disease?
4   Why do caged hens require less food than non-caged hens?
5   Why are the beaks of some chickens trimmed at an early age?

Consumers have taken a very active stance against the living conditions of hens producing eggs. They have decided that hens deserve the opportunity to flap their wings, take dust baths, roost, and experience nesting. They object to massive animal production facilities where wire cages cause birds to suffer. Some people also object to the practices used in raising broilers – chickens raised only for the meat. Large numbers of birds are housed together in barns with poor ventilation. The birds walk on a floor covered with excreta and spilled grain.

Table 5.1 identifies the major issues connected to public concerns about chicken production practices. As governments and business firms respond to these concerns, evidence suggests that some members of the public need more information so they can make choices consistent with chickens' total well-being.

Practices used for raising hens and broilers have developed to reduce the costs of production. Consumers have sought low-cost eggs and poultry products. American producers have responded with practices that lower prices. Large numbers of chickens housed together is economical. It is cheaper to have large numbers of birds in one barn rather than spread out over larger areas at multiple facilities.

Egg prices and labeling

A major concern of many people is whether egg-laying chickens have enough space, which will be covered in a later chapter. This concern has led many retailers to differentiate four categories of eggs: regular (caged), cage-free, free-range, and organic. Consumers who want eggs from cage-free and free-range hens can select these products, and generally pay more for their eggs. However, a majority of consumers simply want low-cost eggs, which are most often from caged hens.

Table 5.1 Major concerns about the treatment of chickens raised for eggs and meat
          products

Practice	Concerns	Responses
Caged hens	Hens cannot express themselves	Regulations on space
Free-range eggs	Hens exposed to more diseases	Producer incentives to keep hens healthy
Beak trimming	Suffering when too much removed	Limitations on methods and amounts removed
Induced molting	Suffering of hens forced to molt	Limitations on measures permitted
Confined broilers	Poor air quality in enclosed housing	Producer incentives for healthy birds
Culling males	Unnecessary deaths	Research on early determination of sex

Table 5.2 Retail egg prices related to production practices*

Type of egg	Quantity	Price 2015	Price 2016	Price 2017	Price 2018
Caged – US grade AA	White 12 pack	$2.99	$1.03	$1.04	$1.42
Cage-free	White 12 pack	$2.89	$3.65	$2.79	$2.16
USDA organic	White 12 pack	$4.38	$3.96	$3.73	$3.96

* USDA, 2016, 2018

Yet, the demand for free-range eggs continues to increase, even though they may be twice as expensive (USDA, 2016). Organic eggs may be more than three times as expensive as eggs from caged hens.

Table 5.2 reports retail prices from the US Department of Agriculture (USDA) for three types of eggs relating to their production practices. It may be noted in 2015 a bird flu outbreak in caged flocks in Iowa and other areas led to the destruction of millions of birds. A shortage of eggs from caged hens caused their prices to temporarily exceed the cost of cage-free eggs being produced in other areas.

In 2016, the demand by supermarkets and restaurants for cage-free eggs caused prices to be higher than might be expected. The prices in 2017 are reflective of normal retail prices. In any case, the need for larger housing and more feed means that cage-free eggs are more expensive than eggs from caged birds. However, with more cage-free housing systems being built, prices for cage-free eggs should continue to drop.

Given preferences for better treatment of egg-laying hens, consumers are seeking truthful labels on egg products. In 2014, a coalition of individuals and animal rights organizations brought a lawsuit to force the federal government to require labeling of egg cartons denoting the treatment of hens (*Compassion over Killing vs. Food and Drug Administration*, 2014). The petitioners requested that the Food and Drug Administration (FDA) promulgate regulations requiring that all eggs be labeled as "free range," "cage-free," or "eggs from caged hens."

The FDA objected to labeling for multiple reasons. First, the agency claimed it lacked authority to regulate egg labeling based on animal welfare. Next, the evidence regarding differences in nutritional content and food safety attributable to the use of cages was not definitive. Finally, the agency noted that these labeling requests were not a priority given constraints on the agency's resources. The court agreed with the agency. The federal government does not need to adopt regulations on labels for eggs regarding production practices.

Comparing cage, cage-free, and free-range eggs

Many consumers believe that free-range and cage-free eggs are superior to eggs from caged hens.

Free-range eggs do tend to have greater nutritive value (Radu-Rusu, 2014). However, eggs from caged hens tend to be richer in energy and fats. For the most part, consumer preference for cage-free and free-range eggs is reflected in the freedom of hens to engage in natural behaviors such as roosting, nesting, taking dust baths, and scratching for food. Yet other features related to production practices are important.

Consumers might want to look beyond freedom of expression and consider risk of disease, egg contamination, bird mortality, environmental footprint, and egg freshness (Table 5.3). The consideration of these features reveals some surprising facts. The summary of the positive features that might be expected to accompany each of the three production methods suggests that eggs from caged birds have some advantages (Centner, 2017). Yet, the public remains convinced that the first feature, freedom of movement, is the most important.

Operators of caged hens usually have biosecurity measures to keep their hens free from disease. Cages offer hens a more hygienic environment so that they have a lower risk of parasitic and bacterial disease (Blokhuis et al., 2007). Conversely, free-range hens are exposed to animals, rodents, birds, and insects that can lead to diseases (Gast et al., 2015). Research suggests that free-range flocks have an increased risk of exposure to potential *Salmonella* vectors (Holt et al., 2011).

Table 5.3 Rating positive and negative features accompanying egg production practices

	Caged	Cage-free	Free range
Hens free to express themselves	Negative	Positive	Positive
Decreased risk of disease of hens	Positive	–	Negative
Decreased risk of contaminated eggs	–	Negative	Negative
Reduced hen mortality from pecking	Positive	Negative	Negative
Environmental footprint	Positive	Negative	Negative
Freshness of eggs	–	–	Negative

Eggs produced at cage-free operations may have less complete biosecurity measures that prevent cross-contamination by equipment and personnel. In 2016, the FDA issued a warning letter to a cage-free operator in Missouri. The letter claimed the operator's cage-free facility was contaminated with filth, without sufficient environmental testing or biosecurity measures to control the spread of *Salmonella* enteritis, lacking cleaning or disinfection procedures to control disease, and no proper refrigeration of eggs (Flynn, 2016). Thus, some cage-free facilities experience more contamination problems, including exposure to *Salmonella*.

A recent experiment suggests that bird density is not a significant factor in aggravating some disease infections. Birds in experimental conventional laying cages and enriched colony laying cages had similar rates of infection of *Salmonella* Heidelberg and *Salmonella* Typhimurium (Gast et al., 2017).

Eggs from caged hens have a lower probability of being contaminated (Radu-Rusu, 2014). However, eggs from caged hens are dirtier due to being laid in the same region as the birds' excreta (Ahammed et al., 2014).

The natural behavior of chickens includes pecking weak and small hens, which can lead to a painful death for weak hens and even cannibalism. While this problem occurs whenever hens interact with other hens, more deaths occur in group settings (Blokhuis et al., 2007). This means that cage-free and free-range eggs come from flocks in which more hens experience torture from being pecked to death.

In adopting cage-free and free-range production facilities, additional inputs are needed to produce the eggs desired by consumers. The facilities require larger housing facilities and the production of more grain for food. More resources are required to produce cage-free and free-range eggs leaving a larger environmental footprint and less land for wildlife and forests.

A common belief is that cage-free eggs are fresher. This depends on the amount of time that the eggs take to get to the consumer. Large caged operators have very efficient distribution networks to major supermarkets. Free-range eggs may come from farms that lack the infrastructure for quick delivery and refrigeration. Freshness is more often dependent on how long the eggs are in the retail establishment and the consumer's refrigerator.

While consumers should be able to have choices on the type of egg products they purchase, governmental requirements on space may be disadvantageous. Adopting regulations requiring more space is accompanied by higher hen mortality, increased risks of disease for hens, more land needed to grow grain, and more opportunities for the eggs to become contaminated.

Broilers

Although a majority of the public has not thought about the conditions of the broilers that provide our inexpensive chicken products, some object to customary production practices. Table 5.4 summarizes the issues with ideas on how to improve conditions for birds and humans.

Table 5.4 Improving conditions for the production of broilers

Condition	Problem	Response
Inside air	High ammonia concentrations	Better air circulation
Inside air	Too much particulate matter	Change litter or improve air circulation
Outside air	High pollutant concentrations	Install biofilters, biotrickling filters, and air scrubbers
Litter disposal	Odors	Proper management and incorporation into the soil

For most broiler operations, thousands of birds are raised together in large houses. High densities of birds can contribute to high ammonia concentrations. The enclosures mean the birds never see daylight. They walk around in their feces mixed with feathers, spilled feed, and litter. The ammonia and hydrogen sulfide from their excrement and the dust from litter and feather particles cause the air quality to be bad.

Dust in the air of poultry barns is a problem due to chickens scratching for spilled food and flapping their wings. High concentrations of particulate matter originate from the litter, spilled food, and excreta on the floor. Wood shavings and sawdust are most often used as litter, although rice or peanut hulls may be used in some regions.

Poultry houses have large fans to remove some of the dirty air. However, concentrations of ammonia, hydrogen sulfide, and dust adversely affect the birds by causing stress and increasing the susceptibility of broilers to disease. Particulate pollution is associated with health risks for not only chickens but also workers inside the houses. Research suggests workers suffer a higher prevalence of respiratory symptoms, asthma, chronic bronchitis, and nasal irritation (Senthilselvan et al., 2011).

Removal of the air and litter can raise concentrations of ammonia in the outside air. The high ammonia content may adversely affect neighbors. Research is being conducted to reduce air pollution and odors coming from poultry houses. Biofilters, biotrickling filters, and air scrubbers can capture and treat air pollutants in structures. Land applications of litter need to be timed to minimize the adverse environmental effects. When applying poultry litter to cropland, it should be incorporated into the soil by plowing or cultivation as soon as possible to reduce smells and losses of nitrogen.

Beak trimming

Another chicken production issue is beak trimming. This practice usually involves the removal of one-quarter to one-third of the upper beak. In some cases, a portion of the lower beak is also removed. An unanswered question is whether there is a preferred debeaking method.

Table 5.5 Evaluating practices associated with beak trimming of chickens

Practice	Usage	Comment
No trimming	Broilers produced for meat	No expense or pain
Hot blade trimming	Future egg-laying hens	Human error causes some painful outcomes
Infrared trimming	Future egg-laying hens	Reduces handling stress and poor results
Trimming curved tip	Birds producing organic eggs	Little or no pain

Beak trimming inhibits the ability of birds to peck the feathers of other birds, thereby reducing pecking injuries and death. A second benefit of beak trimming is that it reduces feed waste. Table 5.5 provides a summary of four issues involved with deciding whether to use this practice.

Hot blade trimming and infrared energy treatment are the most common practices employed to trim beaks. Hot-blade beak trimming uses a heated blade to cut and cauterize the beak tissue of both the upper and lower beaks of young chicks. Infrared beak treatment uses a laser to cause the tip of the beak to soften and erode away.

Beak trimming is beneficial for hens laying eggs as it reduces damage caused by feather pecking and cannibalism (Dennis and Cheng, 2012). Laying hens include broiler breeders that provide a majority of the chickens providing meat products. Beak trimming is generally not used for the production of broilers as they are marketed prior to major feather pecking damage.

As an invasive amputation procedure that causes pain, beak trimming has elicited concern. However, most researchers and producers conclude that it is a beneficial procedure because it reduces feather pecking and cannibalism (Henderson et al., 2009). The USDA examined beak trimming and noted it was needed until a better approach to control feather pecking was available. Generally, beak trimming is allowed if performed in a humane manner.

The European Union allows beak trimming prior to ten days of age. However, certain countries prohibit this practice, including Norway, Sweden, Finland, Denmark, and Austria.

Induced molting

Another chicken production issue is induced molting of chicken flocks. After about 12 months of egg production, hens are ready for a break, and their diminished egg production may mean that they are unable to provide a positive economic return. Operators have two choices. The first is to have them slaughtered and secure a new flock. The second is to have them molt and become rejuvenated for another year of egg production. Some producers choose molting. This practice, however, can be controversial as it involves some stress to birds.

Molting occurs naturally in the autumn with shortened daylight. However, since most hens are kept inside, including cage-free hens, molting has to be induced by the operator. While curbing artificial light will help start this process, it can be better managed through their diet.

Induced molting may involve the producer withdrawing feed for 7–14 days, and sometimes water. However, in the United States, the United Egg Producers instituted a policy in 2006 under which only non-feed withdrawal methods are permitted (United Egg Producers, 2016). Hens are provided a special feed source and water is available. The molt program causes the hens to lose their feathers, cease to lay eggs, and lose body weight. Afterwards, they commence laying eggs again.

In protecting animals for farm purposes in the European Union, a Council Directive requires a wholesome diet (European Council, 1998). Any molting practice of depriving hens of feed for several days violates this directive.

Certification as humane

Consumers seek labels on the humane treatment of animals. For chicken products, they rely on three major third-party certification programs. One is the "American Humane Certified™" for products meeting standards established by the American Humane Association. The second is the "Animal Welfare Approved" seal for products meeting the standards established by the Animal Welfare Approved organization. A third is "Certified Humane®" by Humane Farm Animal Care.

Table 5.6 looks at the practices of beak trimming and induced molting to examine eligibility for a humane seal or certification. Differences in qualifying for an appellation help explain why some consumers are confused.

The Animal Welfare Approved seal does not allow beak trimming of laying hens or meat chickens (Animal Welfare Approved, 2015a, 2015b). Conversely, the Humane Farm Animal Care program for laying hens allows the Certified Humane® endorsement to be used on products from chickens in flocks susceptible to outbreaks of cannibalism (Humane Farm Animal Care, 2014). However, the birds' beaks must be trimmed at ten days of age or younger, and severe beak

Table 5.6 Management practices that qualify for an endorsement of being humane

Practice	Eligibility	Label, seal, or designation
Beak trimming	Not allowed	Animal Welfare Approved
Beak trimming	Allowed on very young birds in qualified flocks	Certified Humane®
Beak trimming	Allowed to prevent feather pecking	American Humane Certified™
Induced molting	Not allowed	Animal Welfare Approved
Induced molting	Allowed but no withholding feed and/or water	American Humane Certified™
Induced molting	Allowed with specialized diet	Certified Humane®

trimming is not allowed. The American Humane Certified™ label is permitted on chicken products produced at farms practicing beak trimming (American Humane Association, 2016). Firms purchasing eggs might have contracts that limit the amount of a beak that can be removed or specify the technology that can be used for beak trimming so less pain is inflicted on the birds.

Examining the issue of whether an induced molting program is humane, there are again different results. The American Veterinary Medical Association recognizes induced molting and recommends a carefully monitored and controlled procedure (American Veterinary Medical Association, 2010). The Association's support is based on the extension of the productive life of commercial chicken flocks and reduction of the numbers of new laying hens required for egg production.

The Animal Welfare Approved seal cannot be used on eggs produced by a flock in which molting was induced. The American Humane Certified™ and the Certified Humane® endorsements can be used on eggs where molting was induced without withholding feed and/or water.

While the three programs express slightly different qualifications, having a flock molt is a practice that reduces the need to slaughter hens after one year of production. Having hens lay eggs for two years is beneficial because new birds do not need to be raised for production.

Backyard production

With many opposed to the production of chickens in facilities housing large numbers of birds, some individuals have decided to raise chickens in their own backyards. In both the United States and Europe, people have built coops for small numbers of birds. Simultaneously, there is concern about their welfare, waste, dead birds, annoyances for neighbors, and the control of disease.

The most serious issue is controlling disease. In the absence of surveillance, biosecurity, vaccination, and slaughter measures, opportunities for the transmission of important poultry diseases are enhanced (Karabozhilova et al., 2012). These diseases may originate from wild birds in the area or rodents.

Live poultry may have *Salmonella* germs that can infect persons coming into contact with the birds, their eggs, water and feed dishes, or their droppings. Droppings used as fertilizer can contaminate vegetable crops. In 2017, the US Centers for Disease Control and Prevention (2017) recorded 1,120 outbreaks of human illnesses due to backyard chicken production, with 249 hospitalizations in the United States.

Questions about what to do with the excreta and dead birds from backyard chicken production disclose various practices. Some of the waste and dead birds end up in landfills.

Culling of male chicks

Currently, male chicks are culled immediately after hatching. Various culling methods exist, including suffocation and gasification. The reason for not wanting most male birds is that their meat will not meet the quality standards demanded by consumers, and they cannot lay eggs. Animal rights groups have objected to this practice, leading the poultry industry to think of ways to avoid this practice.

The industry is hoping that it will be able to tell the sex of the chicks while they are still in the egg. One possible technology is in-ovo sexing to enable eggs with male chicks to be identified and culled. Other research is looking for technologies to identify male chicks before they are born, ways to increase the percentage of eggs with female chicks, or ways to alter the meat of male birds so it can be used for human food (Krautwald-Junghanns, 2017).

Chlorine cleansing

Processors of chickens in the United States routinely wash chicken carcasses with a chlorine sanitizing solution to remove bacteria. The dilute solutions pose no risk of harm to humans, unless an error is made in the concentration of chlorine. Other sanitizing agents may also be used. Despite treatments, considerable amounts of poultry are contaminated with poultry-associated *Salmonella* serotypes.

Europeans object to chlorine washes, and the European Union has banned US chicken cleansed with chlorine. The fear is not about chlorine but rather that production and processing in the US might be accompanied by poorer hygiene. One of the concerns involves the use of antibiotics in the production of American chicken. Yet, no conclusive evidence has been offered to show that the use of chlorine sanitization contributes to antibiotic resistance or leads to inferior products (EFSA, 2005). Many are watching to see what happens after Brexit. Will the UK allow chlorine-sanitized poultry products to facilitate a trade agreement with the United States?

Foodwashing facts

1   For identical sized flocks, more cage-free hens perish than caged hens.
2   Free-range chickens have an increased risk of contracting a disease.
3   Beak trimming, performed correctly, is a humane practice as it reduces damage to birds that are pecked by other birds.
4   Induced molting of laying hens extends their ability to lay eggs, thereby reducing the need to raise replacement hens.

References

Ahammed, M., et al. 2014. Comparison of aviary, barn and conventional cage raising of chickens on laying performance and egg quality. *Asian Australasian Journal of Animal Science* 27(8), 1196–1203.

Animal Welfare Approved. 2015a. *Animal Welfare Approved Standards for Laying Hens.* Marion, VA.

Animal Welfare Approved. 2015b. *Animal Welfare Approved Standards for Meat Chickens.* Marion, VA.

American Humane Association. 2016. *Animal Welfare Standards for Laying Hens-Cage-Free.* Washington, DC.

American Veterinary Medical Association. 2010. Welfare policies revised with strategic goal in mind: Induced molting, beak trimming among updated policies. *Journal of the American Veterinary Medical Association* 236(2), 146.

Blokhuis, H.J., et al. 2007. The LayWel project: Welfare implications of changes in production systems for laying hens. *World Poultry Science Journal* 63, 101–114.

Centers for Disease Control and Prevention. 2017. *Multistate Outbreaks of Human Salmonella Infections Linked to Live Poultry in Backyard Flocks.* www.cdc.gov/salmonella/ live-poultry-06-17/index.html.

Centner, T.J. 2017. Differentiating animal products based on production technologies and preventing fraud. *Drake Journal of Agricultural Law* 22(2), 267–291.

Compassion over Killing vs. Food and Drug Administration. 2014. 2014 US Dist. LEXIS 176928 (US District Court, Northern District of California).

Dennis, R.L., Cheng, H.W. 2012. Effects of different infrared beak treatment protocols on chicken welfare and physiology. *Poultry Science* 91, 1499–1505.

EFSA (European Food Safety Authority). 2005. Opinion of the scientific panel on food additives, flavourings, processing aids and materials in contact with food (AFC) on a request from the commission related to treatment of poultry carcasses with chlorine dioxide, acidified sodium chlorite, trisodium phosphate and peroxyacids. *EFSA Journal* 297, 1–27.

European Council. 1998. Directive 98/58/EC of 20 July 1998 concerning the protection of animals kept for farming purposes. *Official Journal* L221, 23–27.

Flynn, D. 2016. Cage-free eggs present food safety challenge in Missouri. *Food Safety News.* March 7.

Gast, R.K., et al. 2015. Persistence of fecal shedding of *Salmonella* enteritidis by experimentally infected laying hens housed in conventional or enriched cages. *Poultry Science* 94, 1650–1656.

Gast, R.K., et al. 2017. Colonization of internal organs by Salmonella serovars Heidelberg and Typhimurium in experimentally infected laying hens housed in enriched colony cages at different stocking densities. *Poultry Science* 96, 1402–1409.

Henderson, S.N., et al. 2009. Comparison of beak-trimming methods on early broiler breeder performance. *Poultry Science* 88, 57–60.

Holt, P.S., et al. 2011. The impact of different housing systems on egg safety and quality. *Poultry Science* 90(1), 251–262.

Humane Farm Animal Care. 2014. *Humane Farm Animal Care Animal Care Standards: January 2014 – Egg Laying Hens.* Herndon, VA. http://certifiedhumane.org/wp-content/ uploads/2015/11/Std14.Layers.5A.pdf.

Karabozhilova, I., et al. 2012. Backyard chicken keeping in the greater London urban area: Welfare status, biosecurity and disease control issue. *British Poultry Science* 53(4), 421–430.

Krautwald-Junghanns, M-E. 2017. Current approaches to avoid the culling of day-old male chicks in the layer industry, with special reference to spectroscopic methods. *Poultry Science* 97, 749–757.

Radu-Rusu, R.M., et al. 2014. Chemical features, cholesterol and energy content of table hen eggs from conventional and alternative farming systems. *South African Journal of Animal Science* 44(1), 33–42.

Senthilselvan, A., et al. 2011. A prospective evaluation of air quality and workers' health in broiler and layer operations. *Occupational and Environmental Medicine* 68(2), 102–107.

USDA (US Department of Agriculture). 2016. *USDA National Retail Report – Shell Egg and Egg Products.* September 2. http://search.ams.usda.gov/mnreports/pywretailegg.pdf.

USDA. 2018. *USDA National Retail Report – Shell Egg and Egg Products.* February 16. www.ams.usda.gov/mnreports/pywretailegg.pdf.

United Egg Producers. 2016. *Animal Husbandry Guidelines for U.S. Egg Laying Flocks.* 2016 ed.

6 Wild and farm-raised seafood

Key questions to consider

1 Why are many Americans eating more seafood products?
2 Why are farm-raised seafood facilities becoming so important in providing desired seafood products?
3 What problems are being noted with imports of farm-raised seafood?
4 Are the fish populations in our oceans being competently managed?

Farm-raised seafood includes the production of freshwater and marine fish, crustaceans, and mollusks. It is the fastest growing food-producing sector, as many people believe that consuming more seafood products can contribute to a healthy lifestyle. Over one-half of the world's produced seafood consists of freshwater fish, but other species provide a variety of products enjoyed by many Americans (Figure 6.1). Since 2014, consumers have eaten more food from farm-raised than wild-caught fish (FAO, 2017b).

Farm-raised seafood is especially important given the biologically unsustainable levels of many wild fish species (Msangi and Batka, 2015). Technological efficiencies and harvesting power of fishing vessels have increased catches with declines of stocks available for reproduction. The United Nations Food and Agriculture Organization has concluded that less than one-fourth of the world's fisheries might provide more fish, but three-quarters cannot provide increased catches. If more fish are needed for food consumption, they will need to be raised at farming facilities.

The most popular seafood species consumed in the United States are shrimp, tuna, and salmon (von Stackelberg et al., 2017). Fish containing omega-3 fatty acids have anti-inflammatory properties and contribute to proper fetal development and healthy aging. Yet, concerns about the use of antibiotics in the production of aquaculture products and high levels of mercury in some fish species recommend caution.

An analysis of farm-raised seafood suggests that current methods and practices are not achieving their potential. More efforts are needed to help producers gain management skills, navigate production risks, and integrate aquaculture with existing farm activities and crops (Msangi and Batka, 2015). Governments,

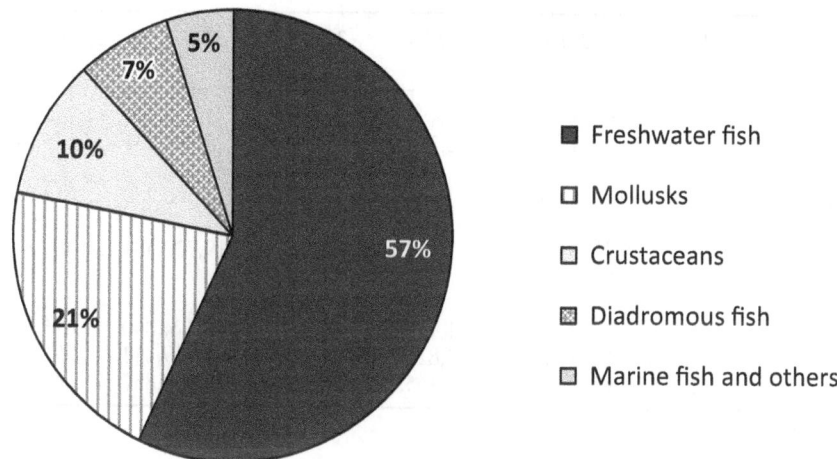

Figure 6.1 World aquaculture production of major species groups*
* FAO, 2017b

marketing firms, and producers can do more to help seafood farming become more productive.

Farm-raised seafood

Although many Americans like fish, our country is not a major producer of aquaculture products. Instead, we import more than 80 percent of the seafood we eat. Asian countries provide the vast majority of the world's farm-raised seafood, as they tend to have lower production costs (Table 6.1). The top ten aquaculture producing countries are China, India, Indonesia, Vietnam, Bangladesh, Norway, Egypt, Chile, Myanmar, and Thailand (FAO, 2017a). These countries provide about 89 percent of the farmed seafood products consumed in the world.

Seafood farming provided more than 76 million tons of food for humans in 2015 and has continued to increase. Approximately 600 species provide aquaculture products, and no single species dominates production (Ottinger et al., 2016). Finfish are the most common and include carp, salmon, catfish, trout, and tilapia. Crustaceans include shrimp, prawn, crabs, and freshwater crayfish. The category of mollusks includes mussels, oysters, and clams.

Many of the most common species of fish are consumed in Asia so may not be familiar to consumers in other parts of the world (Figure 6.2). Five species of carp are raised by many producers. They feed on algae and plants and are easier to grow than other species. Eighty percent of seafood production, mainly freshwater species, comes from small- to medium-scale enterprises (Msangi and Batka,

Table 6.1 Major farm-raised seafood by region*

Region or country	2000	2010	2012	2013	2014	2015
Africa	399,628	1,285,734	1,484,081	1,615,047	1,710,703	1,772,391
Latin America	799,234	1,818,017	2,356,026	2,413,608	2,751,031	2,625,214
North America	584,495	659,040	605,299	597,031	560,980	613,375
Asia excluding China	6,838,995	15,493,933	17,517,188	18,765,117	19,700,581	20,409,817
China	21,527,083	36,741,677	41,114,957	43,555,494	45,474,840	47,615,733
Europe	2,050,689	2,522,707	2,827,124	2,728,580	2,929,242	2,975,159
World	32,418,528	59,027,736	66,465,171	70,247,415	73,729,857	76,641,026

* FAO, 2017b

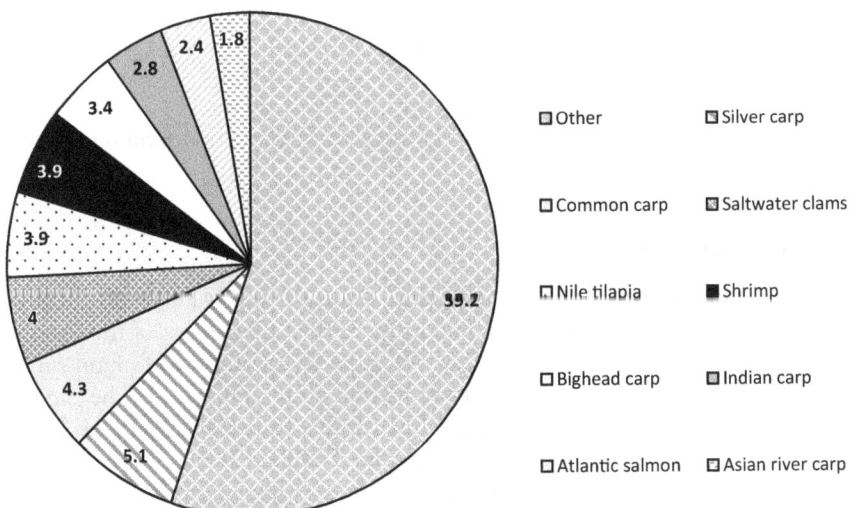

Figure 6.2 Top ten cultured aquatic species by weight (millions of tons)*
* FAO, 2017a

2015). Most marine species are produced at large-scale operations that are capital intense and technologically sophisticated (Miller et al., 2016).

The top seafood imports of the United States are shrimp, tuna, and salmon (NOAA, 2017). Nearly one-half of our Atlantic salmon imports come from Chile, with Canada and Norway also providing significant quantities (USDA, 2018). For shrimp, India provides the greatest volume of imports, followed by Indonesia (USDA, 2018).

Farming systems

Four different culture systems are used for producing farmed seafood: cages, ponds, raceways, and recirculating systems. Cages are systems that confine fish in mesh enclosures, sometimes referred to as net pens. With proper water quality, cage culture can be established in a variety of water bodies, including lakes, ponds, streams, and rivers. Compared to the cost of construction for other aquaculture systems, cage culture is often relatively inexpensive.

Asia is the primary producer of fish products from freshwater ponds. The production practices often integrate fish ponds with agricultural production. For many ponds, animal manure is used as a fertilizer to provide nutrients consumed by the fish (Xiong et al., 2015). The manure may contain antibiotics that were used as growth promoters in the production of pigs and chickens.

Some Asian producers are growing aquaculture species in conjunction with rice farming. Three of the popular products are carp, freshwater shrimp, and freshwater prawns. These fish–rice systems tend to be advantageous because they reduce the need for chemical inputs. The fish can eat insects and weeds, thereby diminishing the need for pesticide use, and less commercial fertilizer is required to maintain yields.

Raceways depend on continuously running sources of water for raising fish and usually consist of manmade structures. The water is discharged, rather than being reused, and removes wastes that are often associated with diseases in farmed fish. Trout and salmon are the most common fish species raised in raceways in the United States. Due to significant costs for feed and operating, as well as competition from larger and foreign producers, few fish products are raised in raceways and recirculating systems.

Although the harvests from farm-raised seafood continue to increase, many producers struggle to control costs and risks that can adversely affect production. Diseases, parasites, and the lack of management skills challenge producers. In other cases, neighbors or governments may express concerns about outflows of organic waste into the surroundings, the transmission of diseases to wild fish populations, and genetic interactions through fish escapes.

Controlling disease

Parasites, bacteria, and viruses can adversely affect the health of aquaculture species and their environments. The high densities of fish and shellfish in farmed systems mean the negative effects of pathogens may increase over time as their hosts cannot escape infection (Tengs et al., 2017). This means than many producers must control disease-causing organisms by using disinfectants, pesticides, and antibiotics.

Semi-intensive and intensive aquaculture systems often rely on antibiotics mixed with fish food to prevent bacterial infections or treat disease outbreaks. Antibiotics can also promote the growth and improve the production of some aquatic products. Excessive uses of antibiotics can lead to the contamination of surface waters as well as excessive concentrations in aquatic products. The use of antibiotics may be contributing to antibiotic resistance.

The Food and Drug Administration (FDA) regulates drugs used in fish farming in the United States. It has approved several immersion products and medicated feeds. More important, we can assume that few unauthorized products are used due to monitoring and enforcement activities under US law. However, many of our seafood products come from countries that lack the enforcement of regulations governing the use of antibiotics (Chuah et al., 2017). Some producers use products without the proper identification of ingredients. In other cases, producers devise their own disease treatment programs.

Salmon production seems to be particularly challenging given the diseases that have developed at production facilities. Sea lice infestations are a major problem, and pesticides are being employed to control this pest. Some producers are using "cleaner fish" to peck sea lice from the skin of farmed salmon, which reduce the need for using pesticides. At salmon farms, vaccines may be used to control diseases, and this practice reduces the use of antibiotics.

Shrimp farming is also subject to disease outbreaks, with acute hepatopancreatic necrosis disease constituting a major problem. The careless movement of living shrimp and use of live and fresh natural feed has enabled the disease to spread to new countries and continents. The industry is responding by urging the cultivation of pathogen-free stocks in biosecure environments (Thitamadee et al., 2016).

Human health

Fish is a nutrient-dense animal-sourced food that is often a part of a healthy diet. It provides micronutrients such as iodine, potassium, selenium, B vitamins, and vitamin D. Many seafood species contain omega-3 fatty acids, which support childhood development and are linked to decreased incidences of allergies in infants. Omega-3 is also associated with improved cardiovascular function in terms of anti-inflammatory properties, peripheral arterial disease, and reduced major coronary events. It is recommended that pregnant women should have fish at an average of two meals a week because of the benefits of omega-3 fats (FDA, 2004).

Yet, as with all food products, safety issues exist. Fish may be contaminated by methylmercury (mercury) and polychlorinated biphenyls. Given the danger of contamination, each US state has compiled consumption advisories to advise people about the health risks accompanying consumption of locally harvested fish. Statistics suggest that millions of individuals consuming self-caught fish in our country are at risk for adverse health effects due to contaminants (von Stackelberg et al., 2017).

The most troublesome contaminant is mercury, as it is associated with long-term neurocognitive deficits in children and impaired cardiovascular health in adults. The federal government advises that women of childbearing age and young children limit their uptakes of fish products containing mercury.

Most of our intake of mercury comes from estuarine and marine seafood, especially from the consumption of long-lived marine fish, including shark, swordfish, king mackerel, and marlin (FDA, 2004). Americans consume large quantities of tuna and shrimp, and approximately 48 percent of our mercury intake comes from these seafood products (Sunderland et al., 2018). Recommendations that the consumption of wild fish is a healthy practice fail to consider mercury contamination.

Antibiotic residues in aquaculture products can pose a health danger, although this is rare. Japan, the European Union, and the United States have adopted regulations to preclude imports with unacceptable residues. Therefore, the more troubling aspect of antibiotic usage is that it contributes to antibiotic resistance. Farm-raised seafood facilities serve as reservoirs for antibiotic resistance that create potential risks to human health (Xiong et al., 2015).

Concerns also exist whether seafood imports from various countries in Asia pose threats to the health of Americans. The FDA is in charge of the safety of seafood imports. Imports are subject to compliance with the regulations of the seafood Hazard Analysis and Critical Control Point Program. The FDA conducts product inspections, inspections of importers, sampling, and assessments of country programs. If imports from a country are found to contain residues of unapproved new animal drugs or unsafe food additives, an import alert may be issued. Under an alert, the FDA can detain, without physical examination, all shipments of identified species from the country, except for firms identified as meeting criteria for exclusions from detention.

Environmental problems

Three major harmful effects have been related to species at fish farms: organic waste, transmission of diseases, and genetic interactions by escaping fish. Fish farming is accompanied by the creation of wastes, mostly in the form of uneaten food and feces from the fish. The wastes are deposited in water bodies and stimulate or exacerbate eutrophication. The death of fish and aquatic plants from eutrophication can adversely affect the ecology of an area.

For species in ponds and impoundments, wastes settle to the bottom but are disturbed when the fish are harvested. Any released effluent will denigrate water quality. Wastes accompanying fish species raised in cages are deposited in the waters and settle to the bottom or are flushed away. Due to the quantities of fish in a given cage, the wastes can have an adverse impact on the associated ecological community.

In some Asian countries, the development of ponds for fish and shrimp have been accompanied by the degradation of coastal lakes, reefs, mangroves, salt marshes, and lagoons. The farms release wastes, nutrients, bacteria, viruses, and other toxic compounds that adversely affect native fish species and contaminate coastal waters. The loss of mangroves alters ecosystems and removes protection against storms, tsunamis, and coastal erosion.

Another issue is whether farmed species can have detrimental effects on wild fish populations. The types of effects include changes to the genetic composition of wild stocks, ecological interactions, and the spread of diseases and parasites (Chang et al., 2014). In 2018, the state of Washington enacted legislation to phase out marine farming of Atlantic salmon due to the nonnative fish escaping and threatening native Pacific salmon runs. Imported carp into the United States have altered the ecology of the Mississippi River and other bodies of water and threaten to devastate the Great Lakes fishing industry.

Resource use

Although environmental issues have been noted, a few additional features about the resources used for farm-raised seafood highlight different concerns. The initial concern is competition for water resources. The use of water for farm-raised seafood diminishes amounts available for other users. In arid areas, and areas with too many demands for water, finite water resources may mean there is not enough water for all users. Furthermore, the wastes released during farm-raised seafood lower the quality of water resources.

A second major issue is the use of fish oil and fishmeal in the production of shrimp, salmon, and trout. It is used as feed because of its high protein levels, excellent essential amino acid balance, and vitamin and mineral content. Most fish used for fishmeal are fish species that could be consumed by wild fish, so their use as fishmeal may detract from wild harvests. This adversely affects local fishermen.

In many of the key areas where fish are being caught for fishmeal, the fisheries are managed to preclude overfishing. Both Chile and Peru set quotas for the amounts of anchovy and sardines that can be caught in their waters. The European Commission has established quotas for various fish stocks in European waters, and the European Union is cooperating with Norway in agreeing to fish quotas in the North Sea. While the management decisions by various officials may not be perfect, decisionmakers are considering the costs of harvesting too many fish. They realize that overfishing leads to lower yields that are accompanied by reductions in economic activities for businesses involved in the fishing industry.

Another issue is the post-harvest losses that are occurring in small-scale fisheries in developing countries (Béné et al., 2016). Spoilage, poor handling, and contamination at small production facilities result in 10 to 20 percent of fish products being unsuitable for human consumption (Hossain et al., 2012). However, in some countries, the losses are higher (FAO, 2010).

A marketing strategy that might be explored further is the certification of aquaculture products. Several organizations exist to certify products including the Marine Stewardship Council, Responsible Fishing Scheme, Dolphin Safe, and Global Aquaculture Alliance. The labels under these programs address different features, including sustainability, hygiene, environmental responsibility, and the use of fewer chemicals. Ecolabeling could be more widely used to support sustainable aquaculture management across the world. These labels could tell consumers about the production practices and whether ecological and socio-economic concerns related to intensive marine finfish aquaculture were considered (Weitzman and Bailey, 2018).

Preservation of wild fish and shellfish stocks

While some health-conscious individuals advocate eating wild fish, this is not realistic given the limited supplies available in the world. The United Nations Food and Agriculture Organization reports that about one half of the world's fish stocks are fully exploited, meaning increased harvests are unlikely. One-fourth of the world's marine fish stocks are over-exploited or depleted. Overexploitation means so many fish are being removed that there is a risk of stock depletion. When stocks are depleted, there are not sufficient adult fish to maintain populations so that fish catches fall below historic levels.

In some areas of the world, countries have developed fisheries management plans to limit catches. In the coastal waters of the United States, requirements delineated in the Magnuson–Stevens Fishery Conservation and Management Act have been important in limiting overfishing. Annual catch limits established by regional fishery management councils have prevented the depletion of some fish populations so that stocks can rebound to levels at which more fish can be harvested. The United States has been successful in rebuilding 44 stocks of fish (NOAA, 2018). The European Union has made progress in replenishing the Atlantic, North Sea, and Baltic Sea fisheries, and two-thirds of the stocks in these areas are being fished sustainably (European Commission, 2018).

However, more work is needed. In the Northwest Atlantic, concerns exist that Atlantic cod, yellowtail flounder, and red hake may be overfished. In the Mediterranean Sea, 93 percent of assessed fish stock are overfished (European Commission, 2016). In the Southeast Atlantic, catch volumes are less than half of what they were 40 years ago, but progress is being made in sustainably managing some species. The Southwest Atlantic and Southern Oceans have also seen significant depletions of selected fish stocks.

Management strategies to enable fish stocks to rebound involve measures that reduce current catches. Catch quotas can cause hardships for local economies dependent on fishing, yet the successful replenishment of stocks eventually results in more profitable fishing endeavors. Given exploitation of ocean seafood species, farm-raised seafood products are expected to become even more important in meeting the world's demand for seafood products (FAO, 2017b).

Governance

Nations and international groups have been active in addressing concerns about maintaining fish populations and managing farm-raised seafood to limit adversities. A range of institutions exist, both public and private, through which interactions between individuals and groups can develop governance structures. Aquaculture governance has centered on the technical upgrading of production to foster improved management standards designed to advance efficiency and reduce negative environmental impacts (Béné et al., 2016).

Individual countries may choose to develop their own multi-level governance, integrated area management, or ecosystem-based management governance provisions. The Norwegian government has been at the forefront of employing governance mechanisms to oversee the preservation of fish stocks and sustainable farm-raised seafood (Sanderson et al., 2014). In addition, and alternatively, international agreements can employ integrated and ecosystem-based management approaches for geographical fisheries. Due to poor environmental regulations and the lack of proper planning and management strategies, fish stocks have been depleted, and aquaculture development has caused serious environmental degradation. More needs to be done to manage fisheries effectively, especially in developing countries and areas of the world without sufficient governance structures.

Three broad elements have been identified for effective shared fisheries (Miller et al., 2016). First, creative mechanisms are needed to make cooperation by each country beneficial. Second, mechanisms are needed to improve the resilience and adaptability of cooperative management arrangements with respect to environmental perturbations. Third, governance must include integration of scientific research and ecosystem monitoring in the management of the shared fisheries. Different stakeholders, philosophies, economic conditions, and political authorities can make gaining agreement on effective management practices difficult.

Recognizing the need for better governance, the Food and Agriculture Organization developed a Blue Growth Initiative to support sustainable fishery and aquaculture policies (FAO, 2014). This initiative can be coordinated with aquaculture practices in developing countries to help support the sustainable development goals of eliminating hunger and ending poverty. Table 6.2 summarizes

Table 6.2 Improving farm-raised seafood in developing countries*

Activity	Challenge	Potential responses
Biosecurity	Poor producers in developing countries have limited capacity to deal with risks	More industry or governmental training and assistance programs
Site limitations	Poor producers not able to select the best sites for production	Help modify facilities to compensate for site deficiencies
Design of facilities	Limited technical knowledge and finances	Industry or governmental programs to encourage better technologies
Sourcing of supplies	Few sources mean high prices	Enhance infrastructures for supplies
Processing	Few buyers mean low prices	Enhance infrastructures for marketing
Retailing	Few buyers mean low prices	Information on certification and branding

* FAO, 2017c

activities and potential responses to help countries make more efficient and effective use of resources in farm-raised seafood.

While the many small production units in developing countries provide significant opportunities for producers and their communities, they are inefficient, often have wasteful post-harvest losses, and fail to maximize the use of resources. Additional training, incorporation of management skills, and coordination of production and marketing activities can improve the efficiency of farm-raised seafood with corresponding benefits to local and national economies.

Summary of concerns

The health and environmental concerns about farm-raised seafood being expressed by citizens and consumers show differences in opinions. There are positive and negative aspects of farmed seafood production. Production is positive because it improves the diets of millions of people in developing countries and relieves poverty. To mitigate the negative environmental aspects of production, governance structures are needed.

Table 6.3 highlights some of the major concerns about farm-raised seafood, with an evaluation of the gravity of each concern and a comparison to whether production is superior to relying on wild seafood products. The gravity attempts to show which concerns might be more significant. While differences of opinion may exist about the gravity of some of the concerns, the major objectives are having more food for the world's population, improving peoples' diets, and the affordability of food. Without farm-raised seafood, prices would rise and fewer people would be able to consume seafood. This would be expected to adversely affect their health.

A major justification for farm-raised seafood is that there are not enough wild fish and shellfish to meet the food demands of the world's population. Consumers want to eat more fish, and fish products are a good source of protein, micronutrients, and omega-3. Fish production in developing countries offers producers, input suppliers, and persons involved in processing and sales gainful economic activities that improve their lives.

Table 6.3 Public concerns about farm-raised seafood contrasted to wild seafood products

Concern	Gravity	Better option
Products contaminated with mercury	Serious	Farm-raised
Products contaminated with polychlorinated biphenyls	Medium	Farm-raised
Antibiotic residues and fostering antibiotic resistance	Medium	Wild
Wastes and environmental damage	Medium	Wild
Damage to indigenous species	Unknown	Wild
Sufficient food production	Serious	Farm-raised
Seafood prices and affordability	Medium	Farm-raised

Foodwashing facts

1   There are not enough wild fish in our streams, lakes, and oceans to meet our demand for seafood products.
2   The United States imports more than 80 percent of its seafood products.
3   Wild fish pose a health hazard as they are more likely to contain mercury.
4   Developing countries need more assistance in managing their aquaculture resources.

References

Béné, C., et al. 2016. Contribution of fisheries and aquaculture to food security and poverty reduction: Assessing the current evidence. *World Development* 79, 177–196.

Chang, B.D., et al. 2014. The development of the salmon aquaculture industry in Southwestern New Brunswick, Bay of Fundy, including steps toward integrated coastal zone management. *Aquaculture Economics & Management* 18, 1–27.

Chuah, L-O., et al. 2017. Antibiotic application and emergence of multiple antibiotic resistance (MAR) in global catfish aquaculture. *Current Environmental Health Reports* 3, 118–127.

European Commission. 2016. *Facts and Figures on the Common Fisheries Policy.* ISSN 1977–3609, Belgium.

European Commission. 2018. *Tackling Overfishing – EU Push for Sustainability Shows Results.*

FAO (Food and Agriculture Organization). 2010. *Post-Harvest Losses in Small-Scale Fisheries: Case Studies in Five Sub-Saharan African Countries.* Technical paper 550. Rome.

FAO. 2014. *Global Blue Growth Initiative and Small Island Developing States.* Rome.

FAO. 2017a. *FAO Yearbook 2015.* Fishery and Aquaculture Statistics. Rome.

FAO. 2017b. *World Aquaculture 2015: A Brief Overview.* FAO Fisheries and Aquaculture Circular No. 1140 FIAA/C1140 (En). Rome.

FAO. 2017c. *The 2030 Agenda and the Sustainable Development Goals: The Challenge for Aquaculture Development and Management.* Rome: Committee on Fisheries.

FDA (Food and Drug Administration). 2004. *What You Need to Know About Mercury in Fish and Shellfish.* EPA-823-R-04–005.

Hossain, M.M., et al. 2012. Quality deterioration of wet fish caught in Mymensingh farm in different steps of distribution channel towards Dhaka. *Journal of the Bangladesh Agriculture University* 10(2), 331–337.

Miller, K.A., et al. 2016. Governing marine fisheries in a changing climate: A game-theoretic perspective. *Canadian Journal of Agricultural Economics* 61, 309–334.

Msangi, S., Batka, M. 2015. The role of fish in global food security. In *The Rise of Aquaculture.* Washington, DC: International Food Policy Research Institute.

NOAA (National Oceanic and Atmospheric Administration). 2017. *Imports and Export of Fishery Products Annual Summary, 2016.*

NOAA. 2018. *2017 Report to Congress on the Status of U.S. Fisheries.*

Ottinger, M., et al. 2016. Aquaculture: Relevance, distribution, impacts and spatial assessments – A review. *Ocean & Coastal Management* 119, 244–266.

Sandersen, H.T., Kvalvlik, I. 2014. Sustainable governance of Norwegian aquaculture and the administrative reform: Dilemmas and challenges. *Coastal Management* 42, 447–463.

Sunderland, E.M., et al. 2018. Decadal changes in the edible supply of seafood and methylmercury exposure in the United States. *Environmental Health Perspectives* 126(1), 1–6.

Tengs, T., et al. 2017. Emerging pathogens in the fish farming industry and sequencing-based pathogen discovery. *Developmental and Comparative Immunology* 75, 109–119.

Thitamadee, S., et al. 2016. Review of current disease threats for cultivated penaeid shrimp in Asia. *Aquaculture* 452, 69–87.

USDA (US Department of Agriculture). 2018. *Aquaculture Data. Economic Research Service.* www.ers.usda.gov/data-products/aquaculture-data/.

von Stackelberg, K., et al. 2017. Results of a national survey of high-frequency fish consumers in the United States. *Environmental Research* 158, 126–136.

Weitzman, J., Bailey, M. 2018. Perceptions of aquaculture ecolabels: A multi-stakeholder approach in Nova Scotia, Canada. *Marine Policy* 87, 12–22.

Xiong, W., et al. 2015. Antibiotics, antibiotic resistance genes, and bacterial community composition in fresh water aquaculture environment in China. *Microbiology Ecology* 70, 425–432.

7 The humane treatment of food animals

Key questions to consider

1 Why do producers castrate male cattle and hogs?
2 Why are horns removed from cattle?
3 Why do hog producers dock their animals' tails?
4 What is a potential problem with animal slaughtering practices?

Consumer demand for low-cost food products has fostered animal production practices centered on prices. In lowering costs, producers of food animals have used management practices that are accompanied by pain, stressful conditions, and suffering. Animal rights groups are expressing concern about whether some of these practices are humane. Reports on objectionable production and slaughter practices are causing the public to advocate limitations on management practices that unnecessarily cause food animals excessive suffering.

The American public has long embraced the humane treatment of pets, especially dogs and cats. Every state has enacted laws offering protection to pets from cruel treatment. However, most of these laws do not apply to food animals. But society's expectations are changing. State legislatures are looking at various livestock management practices to determine whether they are reasonable. If not, legislators can adopt a law to prevent a practice accompanied by suffering.

Europeans have been more proactive in precluding management practices that may inflict pain or suffering on animals. The European Union (EU) adopted a convention in 1976 for the protection of animals kept for farming purposes. Subsequently, five freedoms were recognized for animals being raised for food: (1) freedom from hunger and thirst; (2) freedom from discomfort; (3) freedom from pain, injury, and disease; (4) freedom to express normal behavior; and (5) freedom from fear and distress (EC, 2007).

The Lisbon Treaty, which came into force for the EU in 2009 recognized animals as sentient beings. National governments followed with animal protection laws. Scandinavian countries were particularly supportive of regulatory provisions that would prevent practices that might cause animals to suffer (Viessier et al., 2008).

In the United States, animal rights groups are exposing abusive treatment of farm animals. Undercover videos, sometimes tweaked to exculpate businesses,

Table 7.1 Issues in treating food animals humanely

Practice	Purpose	Implementation qualification
Castration of cattle	Control aggression	Administer pain relief
Castration of hogs	Control boar taint in meat	Perform at an early age
Tail docking of hogs	Reduce tail biting injuries	Perform at an early age
Tail docking of dairy cows	May improve cleanliness	Prohibit the practice
Removing horns	Reduce bruising to other animals	Administer pain relief
Slaughter practices	Proper operation for a fast death	Third party audits

have raised the public's awareness of the pain that accompanies some management practices. Ethical considerations have led citizens and public interest groups to advocate for controls to relieve stress and pain for animals being raised for food.

As more people feel animals should receive better treatment and not be subjected to unnecessary pain or suffering, six practices have garnered attention; castration of cattle and pigs, tail docking of dairy cows and pigs, horn removal from cattle, and inferior slaughter practices. Table 7.1 summarizes the practices, delineates the purpose of each procedure, and identifies an implementation strategy for minimizing discomfort. The humane treatment of chickens and providing animals space also are important and are discussed in other chapters.

Of course, producers of food animals do not engage in these practices to cause animals discomfort. Producers know that they need to keep their animals healthy and comfortable to maximize production of marketable products. Producers use the practices to reduce production costs and improve meat quality. The use of each practice involves a business decision tailored to enhance economic returns. Any regulatory action to change or forbid a practice requires producers to make adjustments to their businesses that can increase costs. Additional production costs will lead to higher meat prices.

Simultaneously, the American Veterinary Medical Association continually reviews production practices and evaluates the merit of practices that inflict pain on animals produced for food. The association balances various criteria, including pain to the animal, physiologic stress, animal health, safety of animals and personnel, and the quality of meat products. As science and technology disclose new information and processes, these production practices are reviewed to determine whether recommendations should change.

Castration of male cattle

A prominent practice that is painful for animals is the castration of male cattle. This practice is considered to be necessary because it leads to carcasses that have higher meat quality and command a higher market price. Castration also reduces aggression, lowers sexual activity, and reduces injuries to people handling cattle.

Table 7.2 Humane certification available for cattle and pigs castrated or tail docked without anesthesia or pain relief

Certification program	Castration		Tail docking	
	Cattle	Pigs	Cattle	Pigs
American Humane Certified™	Under 2 months	Under 7 days	Practice prohibited	Under 7 days
Animal Welfare Approved	Under 2 months	Under 7 days	Practice prohibited	Practice prohibited
Certified Humane®	Requires pain relief	Under 7 days	Practice prohibited	Practice prohibited

The reduction of aggression is particularly important for animals being raised in group settings, which includes most male cattle producing meat products.

For male cattle, three physical methods are surgical castration, Burdizzo clamps, and elastic bands. For each of these methods, the animal will experience stress and pain. The issue is whether the procedure should be accompanied by the use of an anesthetic agent to reduce pain. In most cases, no medication is administered. The US and EU have similar policies on the castration of cattle.

In the United States, a state's regulatory authority may adopt requirements for castration to forbid practices that involve too much pain. Ohio adopted standards that require consideration of the animal's age and weight, environmental conditions, and facilities available as well as human and animal safety (Ohio Administrative Code, 2016). The regulations do not require an anesthetic agent. However, it is unacceptable to castrate cattle over eight months of age.

Since most male cattle are castrated, the question is whether consumers should be able to learn whether their hamburger or steak came from a steer that was humanely treated during its castration procedure. Three groups in the United States have certification programs that address the humane castration of cattle. Table 7.2 summarizes required pain relief for qualifying for a humane designation.

Two of the three humane certification programs allow qualifying bulls to be castrated without anesthesia (American Humane Association, 2016a; Animal Welfare Approved, 2015a). Conversely, castration must be accompanied with local anesthesia or a nonsteroidal anti-inflammatory drug to reduce pain to qualify for the Certified Humane® designation (Humane Farm Animal Care, 2014a).

Castration of male piglets

Most male piglets in the United States are castrated to control "boar taint," an objectionable odor that is in the meat of some uncastrated boars (Tuyttens et al., 2011). Given the placement of testicles, clamps and bands cannot be used on pigs. Thus, most male pigs undergo surgical castration that is accompanied by stress and pain. The use of an anesthetic agent may be advocated but generally is not used. Producers in the EU are supposed to use an anesthetic agent and agreed to phase out pig castration in 2018 (EC, 2010). However, this was a voluntary agreement and many pigs are still being castrated.

For boars, a procedure to replace surgical castration is immunocastration. This involves the injection of a protein compound. In the United States, Pfizer Animal Health markets Improvest® for this purpose. The compound causes the recipient animal to develop antibodies against its gonadotropin-releasing hormone.

Governments have not been active in addressing suffering of pigs during castration. One exception is the humane animal treatment regulation developed by Ohio. It is unacceptable to castrate swine over 75 pounds (Ohio Administrative Code, 2016). The three major humane certification programs do not require the use of an anesthetic agent (American Humane Association, 2016b; Animal Welfare Approved, 2015b; Humane Farm Animal Care, 2014b).

Tail docking of dairy cows

Some dairy farms adopt the practice of docking the tails of dairy cows to prevent the tail from becoming soiled by urine and manure. The bottom part of the cow's tail is amputated and the procedure causes pain. The issue is whether tail docking is a humane practice.

In addition to pain, tail docking also interferes with the cow's ability to dislodge biting flies (American Veterinary Medical Association, 2010). Moreover, research suggests there is no significant improvement in cow cleanliness or milk quality that can be attributed to tail docking (Schriener and Ruegg, 2002). Thus, most people and the American Veterinary Medical Association oppose tail docking of cows. Tail docking of dairy cows is banned in several European countries.

The most common method employed for docking is to use elastrator banding on calves near weaning age (American Veterinary Medical Association, 2010). Due to the band, the bottom portion of the tail detaches from the animal. This procedure takes three to seven weeks. Docking causes some pain but it is not severe.

The California legislature enacted a law in 2009 that prohibits tail docking of dairy cows in California, except in emergency situations (California Penal Code, 2016). Another state that has addressed tail docking is Ohio. Tail docking can only be performed by a licensed veterinarian if the procedure is determined to be medically necessary.

In New Jersey, judicial action led to the prohibition of tail docking. In a lawsuit commenced by the New Jersey Society for the Prevention of Cruelty to Animals, the state's department of agriculture's decision to allow tail docking was challenged (*New Jersey Society for the Prevention of Cruelty to Animals vs New Jersey Department of Agriculture*, 2008). A New Jersey court concluded that in the absence of any support for the practice to offset the pain it causes to animals, tail docking could not be considered to be humane. Subsequently, the state regulation was revised to allow tail docking only upon determination by a veterinarian for individual animals (New Jersey Administrative Code, 2018).

Tail docking of pigs

Many producers routinely dock the tails of piglets to reduce tail biting behavior. Although tail biting is an abnormal behavior, it occurs when pigs are gown for fattening in confined pig facilities. Some feel it is because pigs are unable to perform natural behaviors. Factors that seem to foster tail biting are the floor type, stocking density, ventilation, nutrition, gender, genetics, tail length, and health status. Biting can be abated by providing pigs with straw or another rooting material.

Tail docking is nearly always performed a few days after birth, often at the same time as the castration of male pigs. Producers use this practice to prevent other pigs from chewing on intact tails. However, tail biting also occurs to pigs with docked tails.

Public opposition to tail docking centers on the pain it causes the young pig. Moreover, it is unnatural to remove a tail. As shown in Table 7.2, tail docking is prohibited for the Animal Welfare Approved and Certified Humane® appellations. Pigs with tails docked before they are more than seven days old qualify for the American Humane Certified™ appellation.

Routine tail docking of pigs in the European Union is prohibited. Producers provide enriched facilities with straw to diminish biting. Enforcement of the prohibition varies as some producers claim tail docking is needed to prevent injuries.

Removing horns from cattle

Several benefits accrue to producers who use techniques to eliminate horns on cattle. The most important is decreased bruising of carcasses that leads to discounted carcass prices (Neeley et al., 2014). This means there is a financial incentive to dehorn cattle. There is also a reduction in trauma caused by dominant and aggressive animals, decreased risk of injury to personnel, and easier transport of animals. The removal of horns is especially prevalent in the dairy industry, as 94 percent of dairy cows undergo a procedure to prevent growth of horns (Stock et al., 2013).

Three methods of dehorning cattle are commonly used: chemical application of caustic paste, cautery using a hot iron, and amputation. Obviously, any procedure to remove horns will involve pain. This includes acute pain immediately after the procedure and chronic pain which lasts for a longer period of time. The removal of horns may be accomplished by disbudding or by dehorning. The issue is whether products from cattle that have had their horns removed should qualify for a humane appellation.

The least amount of effort at removing horns is to disbud the horns within two months of birth. Disbudding is usually performed by using a caustic paste or the rapid removal of small horns by scoop. These procedures are generally not accompanied by analgesia (Stafford and Mellor, 2011). All three humane certification designations may be used on products from animals that were disbudded using cautery paste when the animal was under seven days of age.

Table 7.3 Limitations on using a humane appellation on products from cattle that are dehorned

Certification program	Limitations on dehorning
American Humane Certified™	Generally prohibited after 30 days of age except for dangerous animals
Animal Welfare Approved	Prohibited except for the tip of the horn with no living tissue
Certified Humane®	Not allowed as a routine procedure on animals older than 6 months

Older cattle present a more difficult situation. Horned cattle include animals that have foraged on open ranges so never were disbudded. The dehorning of animals that are not young is painful so is nearly always accompanied with the use of analgesic drugs to reduce pain. Local anesthetics and sedative-analgesics help in reducing acute pain. Nonsteroidal anti-inflammatory drugs can mitigate inflammation-associated pain. In the European Union, dehorning of older animals may only be performed by a veterinarian or trained personnel. Table 7.3 summarizes the dehorning limitations as they relate to the qualifications of a humane appellation.

Basically, all three certification programs acknowledge that dehorning of mature animals is not an appropriate practice, so products from dehorned animals should not qualify for humane certification. This is consistent with public sentiment. Since the use of disbudding is recognized as a humane practice, the certification programs encourage producers to employ the practices while the animals are young.

Yet, the consequences of not dehorning cattle, including animals born on the open range, might also be considered. Horned animals inflict pain to other animals, causing bruising to carcasses. Is allowing animals to be battered by horned cattle a humane production practice? Are there situations where dehorning animals might constitute a humane response? The humane appellations on beef products fail to acknowledge whether animals suffered from commingling with horned cattle.

Humane slaughter practices

Another animal welfare issue is whether animals slaughtered for their meat products are handled in a manner to avoid pain and suffering. Since the 1990s, Dr. Temple Grandin, a professor at Colorado State University, has championed the ethical treatment of food animals, and her efforts have revolutionized slaughter facilities (Grandin, 2012). There is no reason for animals to unnecessarily suffer at these facilities. Moreover, an unstressed animal will provide better quality meat products.

It may surprise some people, but large restaurant firms, including McDonald's and Wendy's, have led the way in advancing humane slaughtering techniques. They adopted auditing criteria and procedures to determine whether slaughtering plants were meeting recognized humane standards. The European Union requires humane slaughter procedures, but research suggests that slaughter facilities in other countries may need to update their techniques to meet humane standards (Atkinson et al., 2013).

To be sure a slaughter facility is using humane slaughtering standards, third party audits are employed. An inspector at a facility will observe the animals as they enter the facility and are killed. Inspectors watch for animals falling, the use of electric prods, vocalization by animals, and the insensibility of animals when raised to the bleed rail. Any animal that retains sensibility has not been properly killed, possibly due to faulty operation of the stun gun.

The American Humane Certified™ and Animal Welfare Approved appellations may only be used on meat products from slaughter facilities that pass a review. The review program examines pre-slaughter handling, animal stunning, and the killing of animals. Any downed animal must be immediately euthanized in a manner that renders them insensible to pain.

The American Humane Certified™ and Certified Humane® endorsements cannot be used on products from animals from facilities that do not comply with the American Meat Institute's animal handling guidelines. These extensive guidelines were written by Dr. Grandin to prevent suffering of animals as they proceed to slaughter facilities and during their slaughter.

Religious slaughter techniques

An issue being debated is whether some Jewish and Muslim slaughter rituals should be found to be inhumane. Under kosher and halal rituals, animals are conscious when they are killed. The trachea and esophagus of the animal are severed as well as the carotid arteries and jugular vein. Next, the blood is drained, all while the animal is conscious (Delahunty, 2015). Some people feel that in the absence of a stunned, unconscious animal, there is suffering.

In Europe, a number of countries have decided that the humane slaughter of food animals requires forbidding Jewish and Muslim cattle butchering rituals (Delahunty, 2015). Norway, Sweden, Switzerland, and Denmark have banned kosher and halal slaughter rituals, and similar prohibitions are being debated in other European countries. However, since not all Muslim authorities follow the requirement that animals must be conscious when killed, in some countries the animals can be stunned before they are killed and meet halal requirements.

In the United States, the federal Humane Slaughter Act delineates two methods for humane slaughtering. The first uses a stun gun while the second allows

> slaughtering in accordance with the ritual requirements of the Jewish faith or any other religious faith that prescribes a method of slaughter whereby

the animal suffers loss of consciousness by anemia of the brain caused by the simultaneous and instantaneous severance of the carotid arteries.

(US Code, 2016)

Thus, kosher and halal practices meet legal requirements for humane slaughter in the United States.

Providing for humane treatment

To enable consumers to consider the humane treatment of animals when purchasing food products, there are three major options for requiring or precluding practices. We can rely on governmental regulations, use certified programs with labeled products, or rely on marketing firms to source their animal products from animals treated humanely.

Table 7.4 offers a rating of the potential effectiveness of the three options in precluding practices that cause excessive suffering by food animals. Because most state governments have not been proactive in regulating the four listed practices, it is not clear that this option offers a good solution. Yet, animals that suffer or are unhealthy are not profitable, so producers have incentives to employ practices that minimize pain without governmental oversight.

The best option for consumers who champion the humane treatment of animals is to purchase certified products. The three animal welfare certification programs consider castration, disbudding, dehorning, tail docking, and slaughter practices. Assuming certifiers enforce their provisions, consumers have a good assurance that the animals providing products were treated humanely.

The third option of marketing firms overseeing the human treatment of animals poses difficulties in accountability. Some restaurant chains and supermarkets are aware of these welfare issues and require producers and handlers to avoid repugnant practices. Marketers adopting a humane standard can force producers to meet welfare standards, as they can reject nonconforming products. Producers who fail to meet the standard may need to find a buyer who will accept their animals. Yet, in many situations it is difficult to ascertain that welfare conditions were followed. For cattle, the mixing of animals at grow-out and feeding facilities and the absence of records make it difficult to ascertain whether an animal was treated humanely.

Table 7.4 Support for humane animal treatment

Practice	Government regulations	Certification programs	Marketing firms
Castration	Fair	Good	Poor
Removing horns	Poor	Good	Poor
Tail docking	Some	Excellent	Poor
Slaughter practices	Fair	Excellent	Poor

Conclusion

Americans are demanding changes in practices accompanying the production of food animals. Due to the stress and pain that accompany castration, disbudding, dehorning, tail docking, and slaughtering, governmental and voluntary standards are being adopted that address animal welfare issues. The objective is to only allow practices that meet criteria for the humane treatment of food animals.

Of course, it can be argued that there are no definitive formulations of what animal management techniques are consistent with good animal welfare practices. Moreover, the issue does not involve animal cruelty. Rather, the trend involves requiring practices to be performed in a humane manner. Requirements on humane treatment may require producers to desist from former routine animal management practices.

While well-meaning groups have publicized various production practices as inflicting pain on food animals, the public needs to look at the entire process of securing meat products. Some of the practices are employed to prevent animals from battering and hurting others. A vast majority of producers only use practices that enhance the growth and development of healthy animals. Most of the practices that are being used contribute to the quality of meat and dairy products desired by consumers.

The past two decades have shown that consumers can change animal production practices. Whether through governance or via pressure from marketing firms, producers can be forced to eliminate practices that unnecessarily cause animals pain or to suffer. The most prominent change is the use of humane slaughter practices. If sufficient numbers of consumers want a painful animal practice stopped, the industry will find a way to produce the requested products.

Consumers also need to realize that food animals are not pets. Practices and conditions that are appropriate for their dogs and cats should not be imposed on animals being raised for food. While these animals need to be treated humanely, practices such as castration, dehorning, and tail docking can be beneficial for animal well-being.

Interferences with production practices raise food prices. Every increase in food prices exacerbates the suffering of the 17.4 million households in our country that are food insecure. Food-insecure households already spend 26 percent less for food than the typical food-secure household of the same size and composition (Coleman-Jensen et al., 2015). If prices increase, households with very low food security will experience even more hunger because they lack sufficient money for food.

Foodwashing facts

1   The American Veterinary Medical Association continually evaluates the merit of practices used in the production of food animals.

2 Male animals are castrated to control aggression and reduce injuries to other animals and persons handling the animals.
3 The removal of horns on cows, steers, and bulls reduces injuries to other animals and improves meat quality.
4 Proper slaughter techniques reduce stress of animals and improve the quality of their meat products.

References

American Humane Association. 2016a. *Animal Welfare Standards for Beef Cattle*. Washington, DC.

American Humane Association. 2016b. *Animal Welfare Standards for Swine*. Washington, DC.

American Veterinary Medical Association. 2009. *Welfare Implications of Castration of Cattle*. www.avma.org/reference/backgrounders/castration_cattle_bgnd.pdf.

American Veterinary Medical Association. 2010. *Tail Docking of Dairy Cattle*, January 28.

Animal Welfare Approved. 2015a. *Animal Welfare Approved Standards for Beef Cattle and Calves*. Marion, VA.

Animal Welfare Approved. 2015b. *Animal Welfare Approved Standards for Pigs*. Marion, VA.

Atkinson, S., et al. 2013. Assessment of stun quality at commercial slaughter in cattle shot with captive bolt. *Animal Welfare* 22, 473–481.

California Penal Code. 2016. Section 597n.

Coleman-Jensen, A., et al. 2015. *Household Food Security in the United States in 2014*. United States Department of Agriculture, Economic Research Report #194.

Delahunty, R.J. 2015. Does animal welfare Trump religious liberty? The Danish ban on kosher and halal butchering. *San Diego International Law Journal* 16(2), 341–379.

EC (European Commission). 2007. *Animal Welfare*. Health & Consumer Protection Directorate-General.

EC (European Commission). 2010. *European Declaration on Alternatives to Surgical Castration of Pigs*.

Grandin, T. 2012. Developing measures to audit welfare of cattle and pigs at slaughter. *Animal Welfare* 21, 351–356.

Humane Farm Animal Care. 2014a. *Humane Farm Animal Care Animal Care Standards: January 2014 – Beef Cattle*. Herndon, VA.

Humane Farm Animal Care. 2014b. *Humane Farm Animal Care Animal Care Standards: January 2014 – Pigs*. Herndon, VA.

Neely, C.D., et al. 2014. Effects of three dehorning techniques on behavior and wound healing in feedlot cattle. *Journal of Animal Science* 92, 2225–2229.

New Jersey Administrative Code. 2018. Section 2:8–2.6.

New Jersey Society for the Prevention of Cruelty to Animals vs. New Jersey Department of Agriculture. 2008. 955 A.2d 886 (New Jersey Supreme Court).

Ohio Administrative Code. 2016. Chapter 901:12.

Schriener, D.A., Ruegg, P.L. 2002. Effects of tail docking on milk quality and cow cleanliness. *Journal of Dairy Science* 85, 2503–2511.

Stafford, K.J., Mellor, D.J. 2011. Addressing the pain associated with disbudding and dehorning in cattle. *Applied Animal Behaviour Science* 135, 226–231.

Stock, M.L., et al. 2013. Bovine dehorning: Assessing pain and providing analgesic management. *Veterinary Clinics: Food Animal Practice* 29, 103–133.

Tuyttens, F.A.M., et al. 2011. Effect of information provisioning on attitude toward surgical castration of male piglets and alternative strategies for avoiding boar taint. *Research in Veterinary Science* 91, 327–332.

US Code. 2016. Title 7, Section 1902.

Viessier, I., et al. 2008. European approaches to ensure good animal welfare. *Applied Animal Behaviour Science* 113, 279–297.

8   Providing animals
sufficient space

Key questions to consider

1   Why do consumers object to food animals being raised in confined spaces?
2   Why are animals being raised at crowded facilities?
3   Why are pregnant sows confined?
4   Are free-range chickens healthier?
5   Are animals treated humanely during transport?

Consumers' concerns about practices used to produce meat and dairy products include issues of providing adequate space for animals being raised for food. They object to animal production practices that prevent animals from freedom of movement and from expressing themselves. Consumers want cattle to be able to walk around outside and pigs to rut in the soil. They object to massive animal production facilities where confined quarters place animals in unnatural living conditions.

In the European Union, the five freedoms advocated by the Farm Animal Welfare Council have been important in the development of regulations to protect food animals from crowded conditions. These include sufficient space, interactions between animals, freedom of movement, and enriched environments (Veissier et al., 2008). While the directives of the European Union apply in all member countries, they are implemented by individual member states. Some countries have adopted more demanding requirements.

Table 8.1 identifies seven major issues connected to public concerns about the production of food animals. While space is only one of the issues, some consumer groups recognize that if they secure more space for animals, producers may reduce other undesirable practices such as water pollution and the use of antibiotics.

Governments and businesses are responding to the public's concerns. Producers are being forced to eliminate some production practices that severely limit animals' movement, such as crates for veal calves. The Food and Drug Administration adopted a new Veterinary Feed Directive rule to reduce the use of antibiotics. Firms purchasing animals are telling producers they do not want animals that have been fed feed additives or injected with hormones.

Table 8.1 Major issues accompanying the production of food animals

Issue	Animals	Concerns
Excessive confinement	Veal calves and sows	Animals cannot move freely
Space	Cattle, pigs, and chickens	Objections to confined areas
Large confined facilities	Cattle, pigs, and chickens	Antibiotics, animal waste and water pollution
Antibiotics	Cattle, pigs, and chickens	Development of resistant bacteria
Growth hormones	Cattle	Animal and human health
Feed additives	Cattle and pigs	Residues adversely affecting humans
Manure disposal	Cattle, pigs, and chickens	Water pollution from disposing manure
Nuisances	Cattle, pigs, and chickens	Smelly production facilities and manure

Table 8.2 Changes in confinement for veal calves and pigs

Objectionable practice	Comment on the practice	Changes in response to concerns
Veal calves in small stalls	No need for the practice	Group pens after 10 weeks of age
Sows in gestation crates	Individualized management	Changes to group housing
Sows in farrowing crates	Protects piglets from being crushed	Increasing use of group housing

Confined veal calves and sows

With respect to space, two practices used in the production of meat products from large animals have been recognized as especially offensive. The first is raising veal calves in small enclosures and the second is the confinement of sows in gestation and farrowing crates. Table 8.2 identifies the changes in confinement occurring for veal calves and pigs.

Producers had been raising veal calves in stalls (or crates) where the calves had no room to turn around. The condemnation of this practice has forced producers to change their practices. Through consumer pressure and governmental regulations, the veal calf industry has responded with the goal of having 100 percent of milk-fed veal calves in group pens by 2018 (American Veal Association, 2016). The state of Ohio addressed this issue with a mandatory regulation. Starting in 2018, veal calves must be housed in group pens after ten weeks of age (Ohio Administrative Code, 2016). With these developments, confining veal calves is no longer a significant animal welfare issue.

Another confinement practice that has received attention is the confinement of individual sows in gestation and farrowing crates. Crates are objectionable because the sows are not able to walk around and lack social interactions with other pigs. Two categories of crates may be noted. Gestation crates are used prior to the birth of piglets. Farrowing crates are used for the delivery of new piglets and for about three weeks afterwards.

Gestation crates allow individual sows to be fed according to their needs. As sows reach full term, gestation crates prevent an aggressive sow from injuring another sow. However, the public wants sows to have social interaction and an environment that allow pigs to express their natural behavior (Ryan et al., 2015). Objections to gestation crates have led the industry to change to group sow housing. In several states, gestation crates have been banned.

Farrowing crates allow the producer to assist the sow giving birth to multiple piglets. The specialized crates are constructed to keep sows cooler, while piglets have access to a warmer area when not feeding. Farrowing crates reduce the number of piglets that perish due to sows accidentally laying down on top of them (Hales et al., 2014; Condous et al., 2016).

Sows are transferred to farrowing crates about two days before they deliver their offspring. A producer's individualized record for each sow allows calculation of gestation periods. During a sow's confinement, the amount of food the sow eats is recorded. If a sow suddenly reduces her food intake, the producer or veterinarian can check to determine the problem. Farrowing crates mean that all of a sow's urine and waste is deposited away from her feed. The crates help producers keep their sows and piglets healthy.

Research on whether farrowing crates or group housing is the better option for farrowing is mixed. While some research found farrowing crates result in fewer deaths of piglets, other research found comparable death rates in group housing. Various options for farrowing will continue to be debated, with the industry recognizing the public's desire for pigs to be able to express their natural behavior.

Cage-free eggs

Many consumers object to the production of their eggs from chickens that are cramped into tiny wire cages. Pictures on the internet and claims about dead birds and soiled cages have convinced people that the conditions are not humane. Consumers are not only seeking eggs from cage-free chickens, but are also petitioning for legislation to ban small cages. These objections have resulted in the removal of many small cages, and hens are being raised in enclosures that offer more space.

Due to legislation in 1999, conventional cages were banned in the European Union commencing in 2012. These have been replaced with enriched housing systems and aviaries that include raised platforms allowing increased space per hen. Europeans can buy barn, free-range, or organic eggs that did not come from chickens housed in cages.

In 2010, consumers in California were successful in securing the adoption of a law that makes it illegal to sell eggs from chickens that cannot fully spread both wings without touching the side of an enclosure or other egg-laying hens (California Health and Safety Code, 2016). As a result of this legislation, egg producers in California needed to remodel their laying facilities to provide more room for the hens. This required removing the old cages and installing new systems with more space for each hen.

California went further and also enacted a law prohibiting eggs sold for consumption in the state from coming from a farm where hens do not have the amount of space required under California law. The legality of the requirement has been challenged in federal court. The California law seems to violate the Commerce Clause of the US Constitution.

In addition, significant voluntary actions have come from companies selling food products. Several corporate egg buyers and restaurant chains have adopted policies under which they are committed to purchasing cage-free eggs at some point in the future (Hughlett, 2016). The demand for cage-free eggs is leading additional producers to construct cage-free structures for their laying hens.

Free-range chickens

A considerable number of Americans are committed to buying eggs from chickens that are able to run around outside. The US Department of Agriculture (USDA) has defined free range to mean that the animals have access to the out-of-doors. Free range does not mean the birds were raised outdoors or spent most of their days outside. It simply means they do not have to stay inside and are not in cages. Nearly all free-range eggs come from chickens in enclosed areas.

Under US organic guidelines, all organic poultry are free-range birds. In Europe, hens also are able to go outside. However, there are no rules about how often they get outside and some never may take advantage of the access. Free-range systems generally have higher mortality and higher costs than other production systems (Doyon et al., 2016).

While consumers are concerned about differences in eggs produced from caged, cage-free, and free-range hens, they need to realize a few operators from all types of facilities may fall short of maintaining good production practices. It is especially tempting for an operator experiencing financial difficulties to forgo expenses that should be spent to maintain a healthy environment for hens.

The belief that all free-range chickens are happier and healthier than caged hens is not always true. Their freedom of expression is accompanied by risks. Free-range birds have to be concerned about predators, contacting diseases, excessive heat and cold, and excessive pecking by other hens.

Crowded production facilities

Large numbers of dairy cows, cattle, pigs, and chickens are raised together because it is economical. Due to economies of scale, it is cheaper to have a large number of animals at one facility rather than at multiple facilities. Animals do not expend energy walking around searching for food. Through carefully controlled diets, temperatures, and climatic conditions, animals are made comfortable so they gain weight quickly. Confinement allows the use of antibiotics, hormones,

Table 8.3 Positive and negative features of crowded production facilities

Facility practice	Positive	Negative
Large dairies	Computerized records	Manure disposal and odors
Cattle feedlots	Tender beef	Hormones, manure disposal, and odors
Veal calf confinement	Less barn space	Cramped conditions
Sow farrowing crates	Fewer piglet deaths	Cramped conditions
Confined pig facilities	Control of diseases	Antibiotics, manure disposal, and odors
Caged laying hens	Biosecurity and climate control	Cramped conditions

and feed additives. Moreover, for beef cows, confinement at feedlots leads to meat products that are not as yellow and are more tender (Dunne et al., 2009).

However, confinement practices are accompanied by problems, and the conditions at some of these facilities are not pleasant. The animals are prevented from expressing their natural behavior. Confinement practices and crowded conditions can lead aggressive animals to bully others. Table 8.3 delineates some of the positive and negative features of the crowded conditions.

The major problems are the disposal of animal manure, odors, and the lack of space for animals to move around. Yet, some of the more egregious problems have been addressed. Furthermore, there are advantages to animals being raised in confined facilities. Less feed is needed, so less land is required for the production of meat products.

While people object to raising animals in confined conditions, animals may also suffer at farms where animals are not confined. Skulls in desert settings attest that animals on the range can be stressed finding food and water. A few years ago, hundreds of cattle in South Dakota died from being caught on the range in a blizzard.

Although the profitable production of animals requires them to be healthy, the crowded conditions are contrary to views of how food animals should be raised. People want food animals to have more space so they are comfortable. Consumers are expressing preferences for products from animals that were able to live in a manner consistent with their species. Some consumers refuse to buy meat and eggs if the animals were kept in small crates or cages.

Supermarkets and restaurants respond to consumer preferences. In some cases, state legislatures may respond. Through voter referenda, legislative enactments, and contracts from processing firms, producers are being prohibited from using cages, crates, and confined spaces (Centner, 2010).

Concentrated animal feeding operations

In addition to crowded facilities, the public's concerns over space are expressed by opposition to large concentrated animal feeding operations (CAFOs) that have become prevalent over the past 30 years. Livestock producers have aggregated large numbers of dairy cows, beef cattle, pigs, and chickens in enclosed spaces. Food animals are concentrated at individual farms and in locations with convenient transport to slaughter facilities.

CAFOs house approximately three-fourths of the beef cattle, hogs, and layer hens in our country (US Government Accountability Office, 2008). It is estimated that more than 60 percent of our hogs are at facilities with more than 5,000 hogs (Environmental Protection Agency, 2016).

Table 8.4 lists the major objections to CAFOs, identifies potential responses, and comments on the consequence or status of the responses. Because CAFOs have low production costs, they can deliver low-cost food products. Yet, living near one of these facilities can be challenging. Citizens will continue to request governments adopt regulations that curtail objectionable practices at CAFOs.

All animals at CAFOs are confined. An issue is the amount of space given to animals and their ability to express themselves. Due to the concentrated conditions, oftentimes the animals live on manure-covered surfaces. The large numbers of animals denigrate air quality. Removing manure and disposing of animal waste can contribute to water pollution.

Animal rights groups have claimed farm animals at CAFOs are stressed, diseases are rampant, and that no one cares about the animals. For most CAFOs, this is not true. While conditions may be less than optimal, producers are in the business of raising animals for a profit. Producers want animals to gain weight and stay healthy; otherwise they lose money.

While animals at CAFOs may not be comfortable, their degree of stress is generally moderate. When animals are over stressed, they do not gain weight as fast and are more likely to become ill. Producers with sick or diseased animals are penalized by animals' slow rates of weight gain and additional costs. These producers will go out of business.

The assumption that the production of food animals at CAFOs involves no oversight of animals is also not accurate. Skilled personnel, computerized record systems, and biosecurity measures at these operations monitor animals. Sick

Table 8.4 Public objections to CAFOs, responses, and the status of the responses

Problem	Responses	Status
Confinement of animals	Limitations on the size of CAFOs	No action
Crowded conditions	Limit stocking densities	Marketing actions
Manure disposal	Regulate amounts that can be spread on fields	Regulatory controls
Smells from manure	Regulate the time of application to fields	Regulatory controls
Potential water pollution	Federal and state permitting systems	Regulatory controls
Polluted air near CAFOs	Regulate pollution releases	Minimal regulation

animals can be identified and a veterinarian is on call. There is an incentive to treat sick animals immediately so they do not infect other animals. Any producer who has a major disease outbreak suffers severe financial consequences.

On many dairy farms, computerized records allow the producer to track an animal's feeding pattern and daily milk production so that an illness is recognized quickly (Dairy One, 2016). Any decrease in the amount of milk produced is noticed so that the cow can be further examined and receive medical attention if needed.

Cameras are also being installed at feedlots and other locations for security as well as to record the welfare of the animals. Owners of CAFOs cannot afford to lose animals and so adopt measures to reduce the risk of disease, mistreatment of animals, accidental animal deaths, and theft. Aggregations of animals at CAFOs facilitate expenditures of funds for equipment that can monitor the well-being of animals.

While consumers dislike the crowding of animals at CAFOs, what they mainly object to is bigness. CAFOs are the Walmarts of agriculture. Why can't agricultural production return to the family farms like books describe from the early part of the twentieth century? The answer is production costs. Because most consumers want inexpensive foods, the industry has cut costs by raising animals at CAFOs. Moreover, for the most part, food animals are healthier today than they were 100 years ago.

However, the lack of space for animals at CAFOs is related to challenges in disposing manure. Poor practices and cost-cutting measures may lead to water pollution. Concerns about this issue continue to form a basis for reviling CAFOs.

Water pollution

Aggregations of large numbers of animals at one location create environmental problems. Many members of the public are convinced that CAFOs pollute nearby surface waters with their manure. While this may have been true in the 1990s, it is not clear that CAFOs are significant sources of water pollution today.

During the past 20 years, the Environmental Protection Agency has revised federal regulations governing CAFOs to markedly reduce water pollution. Under the US Clean Water Act, Large CAFOs must have a National Pollutant Discharge Elimination System permit before they can discharge any pollutants into surface waters. This is the same type of permit required for sewage treatment plants. No pollutants are allowed to enter surface waters from a CAFO's physical facilities, including barns, feedlots, and animal waste lagoons.

The permit requirements for CAFOs have drastically reduced amounts of pollutants entering surface waters. In fact, the success of the permitting system has caused nonpoint-source pollution to become a more notable problem. Sources of nonpoint-source pollution include land cultivation, construction sites, erosion, animals drinking out of streams, and manure spread on fields.

The handling of manure from CAFOs is controversial. Although the use of manure as a fertilizer for crop production is encouraged, improper applications can result in unacceptable pollution. Federal and state regulations establish requirements for manure applications to prevent pollution, but lapses by producers and minimal oversight by some state regulators allow pollutants to enter streams and surface waters.

Under federal and state regulations, owners and managers of CAFOs are required to calculate the nutrient needs for each field for crop growth before they apply manure. For applications to fields, producers need nutrient management plans that minimize phosphorus and nitrogen transport from fields to surface waters (Centner and Newton, 2011). Whenever a rain event is predicted, producers cannot apply manure as it could cause pollutants to enter surface waters. If these regulations are followed, CAFO manure is not a significant environmental problem.

Under the regulations, inappropriate and negligent manure applications violate federal and state law. Polluters may be fined up to $25,000 per day for a violation of the federal regulations. The threat of fines means producers generally use care to follow the regulations.

Thus, if water pollution from CAFOs is a problem, it is probably due to the failure of state governments to prosecute offenders. Many states have cut their budgets for environmental enforcement, so they lack sufficient personnel to enforce water pollution regulations. Inspection agencies need personnel to inspect suspected violators, assist violators in coming into compliance, and bring enforcement actions.

Furthermore, it is not clear that spreading animals out at more facilities would reduce water pollution. Farms that are not CAFOs do not need to secure permits that include nutrient management plans. In the absence of required plans, producers may be unaware of the nitrogen and phosphorus content of their manure, may not keep records of manure application, may over-apply manure, and may cause more pollution per animal compared with large producers (Hassinger et al., 2000). Producers who raise animals in pastures may not prevent animals from defecating in surface waters. Food animals raised outdoors in pastures and on the open range are polluting surface waters.

Space during transport

All cattle, hogs, and sheep are transported by road or rail. They may be moved in order to be raised in a different location, to feedlots for fattening, and to the slaughter facility. Concerns exist about the welfare of animals while they are being transported. This includes not only the journey on the vehicle, but also the assembly of animals, loading and unloading them onto vehicles, and penning.

The development of specialized production facilities, feedlots, and slaughterhouses means that most animals are transported more than once during their lives. The conditions during these various aspects of transport tend to be stressful for the animals. They also can lead to injuries, diminished weight gain, and damages to the quality of meat products. Since most animals are moved via transport on roads and highways, aspects of this form of transport will be considered.

Three major concerns are transporting unfit animals, overloading, and conditions during excessive transport distances (Schwartzkoph-Genswein and Grandin, 2014). Producers should not attempt to transport sick or emaciated animals, as they might not survive. For cattle, precaution is advisable to reduce the risk of contracting bovine respiratory disease, also known as shipping fever, caused by stress. Affected animals will have poor growth, lower meat quality, and the disease may lead to death (Schwartzkoph-Genswein and Grandin, 2014).

On vehicles, several concerns are important. First, an economic incentive encourages overcrowding of food animals on a transport vehicle to reduce transport costs. Crowded transport conditions are more stressful and can lead to reduced carcass quality accompanied by lower sales prices. However, despite concerns about the humane transport of animals, the loading densities of animals are not regulated by the federal government. Rather, recommended industry standards exist. Each person shipping animals considers risks of having animals lose weight or die in determining transport densities. Federal rules do require care to be used to protect livestock from extreme differences in temperature.

The federal government did pass the "Twenty-Eight Hour Law" that requires a period of unloading for animals being transported for long periods of time (US Code, 2016, tit. 49). If the expected travel time is more than 28 hours, the animals must be unloaded so they can move around. For animals arriving at slaughterhouses, USDA inspection personnel should ask managing personnel whether the animals had rest, food, and water in the past 28 hours (USDA, 2011).

In addition, the USDA has regulations on enclosures used for the feeding, watering, and resting of livestock during transit (US Code of Federal Regulations, 2018). The enclosures should have sufficient space for all the livestock to lie down at the same time. There are distinct requirements for water for different animal species.

Foodwashing facts

1   Consumer demand for products from animals having sufficient space has improved animals' living conditions.
2   More baby pigs perish when sows are not confined.
3   Free-range chicken systems tend to have higher mortality rates.
4   Manure from CAFOs spread on crop land are governed by detailed rules to minimize pollutant discharges to surface waters.

References

American Veal Association. 2016. *Animal Care & Housing*. Gladstone, Missouri. www.americanveal.com/animal-care-housing/.
California Health and Safety Code. 2016. Sections 25991–25996.
Centner, T.J. 2010. Limitations on the confinement of food animals in the United States. *Journal of Agricultural and Environmental Ethics* 23, 469–486.

Centner, T.J., Newton, G.L. 2011. Reducing concentrated animal feeding operations permitting requirements. *Journal of Animal Science* 89(12), 4364–4369.

Condous, P.C., et al. 2016. Reducing sow confinement during farrowing and in early lactation increases piglet mortality. *Journal of Animal Science* 94(7), 3022–3029.

Dairy One. 2016. *Herd Management Software*. Ithaca, New York. http://dairyone.com/agricultural-management-resources/herd-management-software/.

Dunne, P.G., et al. 2009. Colour of bovine subcutaneous adipose tissue: A review of contributory factors, associations with carcass and meat quality and its potential utility in authentication of dietary history. *Meat Science* 81(1), 28–45.

Doyon, M., et al. 2016. Consumer preferences for improved hen housing: Is a cage a cage? *Canadian Journal of Agricultural Economics* 64, 739–751.

Environmental Protection Agency. 2016. Ag 101. www.epa.gov/sites/production/files/2015-07/documents/ag_101_agriculture_us_epa_0.pdf469–486.

Hales, J., et al. 2014. Higher preweaning mortality in free farrowing pens compared with farrowing crates in three commercial pig farms. *Animal* 8(1), 113–120.

Hassinger, W.J. II, et al. 2000. Nutrient management practices among swine operations of various sizes. *Journal of the American Veterinary Medicine Association* 217, 1526–1530.

Hughlett, M. 2016. The push to kick the cage. *Star Tribune* (Minneapolis, MN), October 19, 2015, p. 1D.

Ohio Administrative Code. 2016. Chapter 901:12.

Ryan, E.B., et al. 2015. Public attitudes to housing systems for pregnant pigs. *PLoS One* 10(11), e0141878.

Schwartzkoph-Genswein, K., Grandin, T. 2014. Cattle transport by road. In *Livestock Handling and Transport*, Temple Grandin (ed.), 4th ed., pp. 143–173. Croydon, UK: CAB International.

US Code. 2016. Title 49, Section 80502.

US Code of Federal Regulations, 2018. Title 9, Sections 89.3, 89.5.

USDA (US Department of Agriculture). 2011. *Humane Handling and Slaughter of Livestock Directive*. Food Safety and Inspection Service.

US Government Accountability Office. 2008. *Concentrated Animal Feeding Operations: EPA Needs More Information and a Clearly Defined Strategy to Protect Air and Water Quality from Pollutants of Concern*. Washington, DC. www.gao.gov/new.items/d08944.pdf8.

Viessier, I., et al. 2008. European approaches to ensure good animal welfare. *Applied Animal Behaviour Science* 113, 279–297.

Part III

Consumer information on inputs

9 Objecting to antibiotics

Key questions to consider

1 Why are people concerned about the use of antibiotics in food animal production?
2 Why do so many producers use antibiotics?
3 Over the past few years, have producers reduced their usage?
4 Should we forbid usage of antibiotics in agriculture?

Antibiotics have long been recognized as miracle drugs. They are used to treat bacterial infections and can prevent more serious complications such as cellulitis or pneumonia. They are also important in precluding infections that may spread to other people and in speeding up recovery. Yet, antibiotic use in animal production to enhance animal growth has generated significant opposition due to the problem of antibiotic resistance.

Bacteria are continually adapting to antibiotics causing them to become less effective. The medical community has identified antibiotic resistance as a major public health threat (Table 9.1). In the United States, more than two million people are infected with resistant bacteria each year, and 23,000 Americans die of infections involving resistant bacteria that cannot be successfully treated with antibiotics (Centers for Disease Control and Prevention, 2013). An estimated 700,000 deaths occur around the world that are related to antibiotic-resistant bacteria (O'Neill, 2014).

The Centers for Disease Control and Prevention (2013) considers antimicrobial resistance to be one of the nation's most serious health threats. Given the costs of illnesses and deaths related to resistance, President Obama issued an executive order in 2014 calling for the implementation of measures to reduce the emergence and spread of antibiotic-resistant bacteria. The President created an advisory council and a task force for combating antibiotic-resistant bacteria. Other provisions of the executive order called for reviewing existing regulations, proposing new regulations, strengthening surveillance efforts, responding to antibiotic-resistant outbreaks, and promoting new antibiotics.

In 2015, the White House issued a National Action Plan for Combating Antibiotic-Resistant Bacteria, enumerating five goals "to strengthen healthcare,

Table 9.1 Yearly statistics on the adverse effects of antibiotic-resistant bacteria

Serious infections in the US	2 million	Centers for Disease Control and Prevention, 2013
Yearly deaths in the US	23,000	Centers for Disease Control and Prevention, 2013
Healthcare costs for the US	$20,000,000	Centers for Disease Control and Prevention, 2013
Social costs in the US	$35,000,000	Centers for Disease Control and Prevention, 2013
Yearly deaths worldwide	700,000	O'Neill, 2014
Anticipated deaths in 2050	10,000,000	O'Neill, 2014

public health, veterinary medicine, agriculture, food safety, and research and manufacturing" (White House, 2015). One of the goals of this action plan concerns limiting the use of antibiotics in food animal production. To slow the emergence of resistance bacteria, the plan sought to eliminate "medically-important antibiotics for growth promotion in food-producing animals and bring other agricultural uses of antibiotics, for treatment, control, and prevention of disease, under veterinary oversight."

The World Health Organization has also been a vocal proponent of limiting unnecessary uses of antibiotics. For more than ten years, the organization has been involved in efforts of classifying medically important antimicrobials as important, highly important, or critically important for human medicine (WHO, 2017). This classification can then be used to guide the use of antibiotics in food-producing animals in a manner that helps preserve the effectiveness of the most medically important antimicrobials.

Therapeutic and nontherapeutic uses

When considering antibiotic uses with animals, most Americans think they are used to treat sick pets or to prevent the spread of bacteria that could infect other animals. This usage is known as therapeutic use: treating known and suspected infections. However, antibiotics are also used for increasing the rate of weight gain or improving feed efficiency. These uses are known as nontherapeutic uses. The federal Food and Drug Administration (FDA) refers to antibiotic usage to increase weight gain or improve feed efficiency as "production uses."

It was estimated that 60–80 percent of antibiotics used in the United States may be administered to food animals for production uses (Center for Food Safety, 2015). This would mean that most of the antibiotics used in the United States are not to treat disease or to prevent disease. Since antibiotic resistance is a problem and resistance is related to usage of antibiotics, the FDA has adopted new regulations that limit uses of many antibiotics used for animal production.

Some of the nontherapeutic antibiotics administered to food animals are the same or very similar to the antibiotics used in humans (Centner, 2017). This

Table 9.2 Critically and highly important antibiotics*

Antibiotic	Animal use	Concerns about the continued use for humans
Tetracyclines	Cattle, swine, poultry	Brucella, Chlamydia ssp., and Rickettsia spp. infections
Macrolides	Cattle, swine, poultry	Limited therapy for Legionella, Campylobacter, and MDR Salmonella and Shigella infections
Aminoglycosides	Swine, poultry	Transmission of Enterococcus spp., Enterobacteriaceae (including Escherichia coli) and Mycobacterium spp. from non-human sources
Sulfonamides	Cattle, swine, poultry	A limited therapy for acute bacterial meningitis, systemic non-typhoidal Salmonella infections, and other infections
Lincosamides	Swine, poultry	Human infection may result from transmission of methicillin-resistant Staphylococcus aureus from non-human sources

* World Health Organization, 2011

raises the concern that bacteria developing resistance to antibiotics administered to food animals will lead to resistant bacteria that infect humans (Table 9.2). Scientists are still researching the possible connections between resistant bacteria in food animals and humans (Aarestrup, 2015).

Recommending reductions in usage

Although scientists do not know the precise connections, it is generally agreed that the more a given antibiotic is used to treat animals, the greater the chance that antibiotic-resistant bacteria will emerge. Several transmission scenarios for resistance may be identified (Table 9.3). Given increases in resistant bacteria, scientists and members of the public want to reduce or eliminate the use of non-therapeutic antibiotics.

Scientists know that antibiotic usage amplifies the presence of resistant bacteria in animals' intestinal tracts (Sun et al., 2014). Research has shown that resistance genes can be transferred between bacterial species (Chenney et al., 2015). People may be exposed to resistant bacteria via the food chain or through contact with infected animals, their feces, or contaminated environments (Chenney et al., 2015).

Recent research has found evidence of potential relationships between uses of animal drugs with resistant bacteria affecting humans (Alba et al., 2015). Livestock can serve as a reservoir for transferring resistance genes to humans directly or through food products (Schmithausen et al., 2015). Livestock-associated

Table 9.3 Documenting the transmission of antibiotic-resistant bacteria

Concern	Health hazard
Unnecessary nontherapeutic usage	Increases gene pool and amplifies the presence of resistant microorganism strains
Transfer to other bacteria	Resistance genes can be transferred between bacterial species
Transfer from animals to humans	Resistant bacteria may be acquired through contact with infected animals, their feces, or contaminated environments
Transfer from animals to humans	Transferable resistance genes through food products
Transfer to farm personnel and dogs	Resistant bacteria transmitted on the farm and to humans

methicillin-resistant *Staphylococcus aureus* may be associated with at least 10 percent of these infections in humans (Cuny et al., 2015).

The unnecessary administration of antibiotics to food animals for nontherapeutic uses has led the medical community and public health officials to recommend regulatory actions to curb nontherapeutic usage. In 2012 and 2013, the FDA issued two guidance documents that established principles for changing the regulatory oversight of antibiotics used in agriculture (FDA, 2012, 2013). In 2015, the FDA adopted new federal regulations on the administration of nontherapeutic antibiotics known as the Veterinary Feed Directive (FDA, 2015).

The Veterinary Feed Directive adopts the principle that the use of medically important antimicrobial drugs in food-producing animals should be limited to those uses that are considered necessary for assuring animal health. This means the Directive addresses the use of nontherapeutic antibiotics by directing that antibiotics should not be administered for growth promotion.

The Veterinary Feed Directive also adopts the principle that the use of medically important antimicrobial drugs in food-producing animals should be limited to those uses that include veterinary oversight or consultation (FDA, 2012). Moreover, the Veterinary Feed Directive caused some over-the-counter drugs to require a written order by a licensed veterinarian before they can be used to treat food animals. This means that persons who are not veterinarians cannot prescribe antibiotic drugs for food animals that are covered by the Directive.

Producer profitability using nontherapeutic antibiotics

Producers administer nontherapeutic antibiotics due to the profits associated with usage. Usage increases the rate of daily weight gain of food animals, allowing them to be marketable in less time (Sneeringer et al., 2015). Usage allows animals to eat less food per unit of weight gain, thereby lowering feed costs. For young animals, usage reduces mortality rates and reduces illnesses. For some

Table 9.4 Summary of producer activities to reduce development of antibiotic-resistant bacteria

Application	Activity or safeguard
Administering to animals	Only when necessary to treat infectious diseases
Usage oversight	Require a licensed veterinarian with a connection to the production facility
Production facilities	Adopt hygienic practices and biosecurity practices that are related to healthier animals
Vaccination	Vaccinate when possible to avoid future antibiotic usage
Animal care	Use sustainable systems and good handling practices

animal species, nontherapeutic antibiotics can enhance artificial insemination success rates in female animals to shorten the time between bearing offspring. For swine, usage can increase the number of live animals in a litter.

Administration of nontherapeutic antibiotics is also associated with reduced production costs. Fewer animals will require individual veterinary care, saving veterinarian costs as well as time of the producers. Sick animals tend to not gain as much weight, so by keeping animals healthy, usage enhances weight gains. Usage also can lower feed costs.

However, producers can be encouraged or forced to reduce their use of non-therapeutic antibiotics. The FDA's Veterinary Feed Directive, adopted in 2015, offers a starting point. Table 9.4 summarizes several ideas to reduce usage.

More detailed provisions are needed so that the use of antibiotics in animal production is only to prevent disease or treat illness. To administer the usage of antibiotics, governments can rely on veterinary oversight with mandatory reporting requirements on the usage of critically and highly important antibiotics. At the farm level, producers can adopt practices and measures to reduce the risk of a disease outbreak and to keep animals healthy. In some cases, vaccination programs can reduce the risk of animals becoming ill. For all production facilities, sustainable systems and good practices can also assist in keeping animals healthy so antibiotics are not needed.

Consumer choices

Many Americans oppose the use of antibiotics in food animals because usage enhances the development of antibiotic-resistant bacteria. While the industry and government attempt to reduce the use of nontherapeutic antibiotics, as consumers we can take further action. We can use our purchasing power to encourage producers to administer fewer antibiotics to food animals.

Some consumers would like to ban all uses of antibiotics in animals. However, this is not a good idea. Antibiotics are important in reducing the suffering of animals from bacterial infections. Owners of animals deserve access to therapeutic antibiotics to treat sick animals.

However, our efforts to reduce antibiotic usage can be initiated with our purchases of meat products. We can seek products that are labeled and tell us that no antibiotics were administered to the animals supplying the products. This labeling program would include a mechanism for oversight to guarantee the truthful labeling of products from animals not administered antibiotics.

Producers will stop using antibiotics if it becomes profitable. This generally means that products produced without antibiotics need to sell at a higher price. This requires a market where products from animals that were not administered antibiotics can be identified and the prices of the products are high enough to make it profitable for producers to discontinue antibiotic usage. Consumers must be willing to pay higher prices so that the price premium encourages producers to refrain from using antibiotics.

Similarly, suppliers will provide products from animals not receiving antibiotics if it is profitable. This will require a labeling program that guarantees products from animals that never received antibiotics can be labeled so consumers have a choice.

Actions by vendors, stores, and restaurants

Vendors, stores, and restaurants are responsive to consumer demands. They have recognized that considerable numbers of consumers seek meat products from animals that were not treated with antibiotics and are working to supply such products.

First, major meat vendors are adopting programs to reduce or eliminate the use of antibiotics in animals providing their products. Much of the vendor action has occurred in the poultry industry due to the widespread use of nontherapeutic antibiotics in raising chickens. This often starts with a vendor contracting with growers to abstain from using antibiotics. Since vendors know which flocks did not receive antibiotics, they can separate the birds and products from others that are connected to the use of antibiotics. This allows the vendor to label products from chickens not receiving antibiotics.

Pilgrim's Pride continues with its goal of curbing antibiotic use. The firm announced that 25 percent of its chicken products would be antibiotic free by 2019 (Bunge, 2015). Perdue Farms, Inc. has taken steps to help producers eliminate antibiotics in producing chickens. The company claims that 95 percent of its chicken production is eligible to be sold under the label "no antibiotics ever" (Charles, 2016).

Tyson Foods is using its "no-antibiotics-ever" label for three of its labels: Tyson Red Label™ brand, Tyson True®Tenderpressed® brand, and select Tyson® (Tyson Foods, 2018). Its organic brand also guarantees no antibiotics were used.

Supermarkets are also aware of consumer demands regarding antibiotics. Many rely on vendors for providing meat products from animals not treated with antibiotics. Others have marketing strategies geared for consumers who consciously seek healthier meat products, with attributes including no antibiotics, no feed additives, and no hormones.

Restaurants are also watching consumer preferences. In 2015, McDonald's announced that it would begin using chickens that are not raised with antibiotics (Strom, 2015). Subway has also moved to source meat products from animals not receiving antibiotics.

Claims that meat products come from animals that were never treated with antibiotics are often difficult to verify. Currently, many firms are simply saying they have instituted programs to reduce uses of antibiotics and identifying projections for the future. With these public relations programs and projections, there are no definitive claims that their products come from animals not receiving antibiotics.

Verification and certification

Given the ease for sellers of meat products to make fraudulent claims about the nonuse of antibiotics, consumers might seek verified or certified products. Product verification involves documentation that claims about products are true. Certification occurs through a third party, often a certifying organization. Certification consists of a guarantee about the truthfulness of a claim about a product.

Currently, meat products cannot be labeled as "antibiotic free" due to inability to verify through an antibiotic-residue test that no antibiotics were administered to the animal (USDA, 2002). However, various verification and certification options are available to inform consumers about the nonuse of antibiotics in animals providing meat products.

For labels touting that no antibiotics were administered to be successful, there must be a mechanism to prevent fraud. This might involve open veterinary records of the producers involved in producing the animals to disclose that the animals were not administered antibiotics and tracing the animals to the finished meat products. State, federal, and private certification programs are available for providing these guarantees.

The certification of whether antibiotics were administered to animals is difficult to oversee and creates opportunities for misconduct and cheating. The provision of information about attributes of animal food products starts with self-reporting by producers. For some labels, producers may be the only source of information on how the animals were raised. For other products, there are reports to the government or certification by a third party.

Substantiation by governments and third parties on how animals were raised enhances the probability of truthful information. For livestock and meat products, an additional level of oversight can be provided by participation in fee-based USDA Quality System Verification Programs (US Code of Federal Regulations, 2018). Under these voluntary programs, officials of the agency conduct on-site assessments and publish a listing of companies that have qualified under a program (US Code of Federal Regulations, 2018; USDA, 2015). In this manner, there is further assurance that the products conform to statements on their labels.

Certification by third parties is also possible, under which certified products bear the certifier's mark on the packaging. These involve the use of an independent group to ascertain that animals were raised in a manner to meet the standards associated with a product's label. Applicants seeking certification submit information to a third party to show they qualify for an appellation. The USDA's organic certification employs third party certification.

There are also other notable programs, including "American Humane Certified," "Food Alliance Certified," and "American Grassfed Certified" that offer consumers verification beyond that offered under other governmental and voluntary labeling. Third party certification involves an added guarantee about the truthfulness of the label.

Future ideas to reduce usage

Scandinavian countries have been successful in eliminating many production uses of antibiotics (Bengtsson and Wierup, 2006). Their programs offer support for efforts in the United States to reduce antibiotic usage in animal production.

While the provisions of the Veterinary Feed Directive are significant, Table 9.5 shows that large quantities of antibiotics used in animal production are simultaneously used to prevent disease as well as promote growth. This allows producers and their veterinarians to decide that the use of an antibiotic is justified to prevent disease when in fact a major purpose is to promote growth. The differentiation between preventing disease and promoting growth will need to be evaluated to determine how to further reduce antibiotic usage.

To achieve significant reductions of antibiotic usage, more definitive regulations may be required. California legislators enacted a law in 2015 with provisions that go beyond the Veterinary Feed Directive regulations to curtail the use of antimicrobial drugs. The objective was to foster insights on links between antimicrobial use patterns in livestock and the development of antimicrobial resistant bacterial infections (Table 9.6). The California provisions show that additional methods are possible to encourage reductions in the use of antibiotics.

California's law prescribes restrictions on uses of antibiotics for preventing disease (California Food and Agriculture Code, 2018, § 14402). Under the law, producers may use medically important antimicrobial drugs only in three situations.

Table 9.5 Classes of antibiotics for disease prevention and feed efficiency/growth promotion

Antibiotic class	Preventing disease			Efficiency and growth		
	Chickens	Cattle	Swine	Chickens	Cattle	Swine
Lincosamides	x		x	x		x
Macrolides	x	x	x	x	x	x
Penicillin			x	x		x
Streptogramins	x	x	x	x	x	x
Tetracyclines	x	x	x	x	x	x

Table 9.6 Features to encourage the reduction in uses of antibiotics in animal production

Feature	Possible new rule
Site visit prior to prescribing a drug	California Code of Regulations, 2015, § 2032.1
Proscribe disease prevention	California Food and Agricultural Code, 2018, § 14402
Producer records	California Food and Agricultural Code, 2018, § 14406
Stewardship guidelines	California Food and Agricultural Code, 2018, § 14404
Best management practices	California Food and Agricultural Code, 2018, § 14404
Penalties for violations	California Food and Agricultural Code, 2018, § 14408
Educational rehabilitation for violators	California Food and Agricultural Code, 2018, § 14408

First, when uses are necessary to treat a disease or infection. Second, when uses are necessary to control the spread of a disease or infection. Third, when uses are necessary in relation to surgery or a medical procedure. In all other cases, medically important antimicrobial drugs shall not be administered in a regular pattern.

The California law responds to the contention that the Veterinary Feed Directive is too expansive in allowing antibiotics to be used for producing food animals. The California provisions limit preventive antibiotic usage to prophylactic drugs addressing an elevated risk of contracting a particular disease or infection.

In addition, the California Department of Food and Agriculture has been directed to more proactively regulate the usage of antibiotics. The Department is obliged to develop antimicrobial stewardship guidelines and best management practices that can result in phasing out some uses of antibiotics. One example would be delineating how antibiotics can be administered for shorter lengths of time.

The California law also includes civil penalties. Persons violating the law can incur fines and can be required to attend an educational program on the judicious use of medically important antimicrobial drugs.

Foodwashing facts

1   Resistance to antibiotics is a worldwide problem, with thousands of people dying every year due to the lack of effective medical procedures.
2   A majority of the antibiotics used in the United States are administered to food animals.
3   Some antibiotics administered to food animals are also used to treat humans.
4   Many of the antibiotics administered to food animals improve feed efficiency or enhance growth but are not medically necessary.

References

Aarestrup, F.M. 2015. The livestock reservoir for antimicrobial resistance: A personal view on changing patterns of risks, effects of interventions and the way forward. *Philosophical Transactions of the Royal Society B*, 370(1670), 1–13.

Alba, P., et al. 2015. Livestock-associated methicillin resistant and methicillin susceptible *Staphylococcus aureus* sequence type (CC)1 in European farmed animals: High genetic relatedness of isolates from Italian cattle herds and humans. *PLoS One* DOI:10.1371/journal.pone.0137143.

Bengtsson, B., Wierup, M. 2006. Antimicrobial resistance in Scandinavia after ban of antimicrobial growth promoters. *Animal Biotechnology* 17, 147–156.

Bunge, J. 2015. Pilgrim's expects 25% of its chicken will be antibiotic-free by 2019: Company is also working to end chicken operations' use of antibiotics needed to fight human illnesses. *Wall Street Journal* (online), April 20.

California Code of Regulations. 2015. Section 2032.1.

California Food and Agriculture Code, 2018. Sections 14400 to 14408.

Center for Food Safety. 2015. *America's Secret Animal Drug Problem: How Lack of Transparency Is Endangering Human Health and Animal Welfare*, September.

Centers for Disease Control and Prevention. 2013. *Antibiotic Resistance Threats in the United States*. Atlanta. www.cdc.gov/drugresistance/pdf/ar-threats-2013-508.pdf.

Charles, D. 2016. *Perdue Goes (Almost) Antibiotic-Free. The Salt*. NPR.

Centner, T.J. 2017. Differentiating animal products based on production technologies and preventing fraud. *Drake Journal of Agricultural Law* 22(2), 267–291. www.npr.org/sections/thesalt/2016/10/07/497033243/perdue-goes-almost-antibiotic-free

Chenney, T.E.A., et al. 2015. Cross-sectional survey of antibiotic resistance in *Escherichia coli* isolated from diseased farm livestock in England and Wales. *Epidemiology and Infection* 143, 2653–2659.

Cuny, C., et al. 2015. Livestock-associated MRSA: The impact on humans. *Antibiotics* 4, 521–543.

FDA (Food and Drug Administration). 2012. *Guidance for Industry: The Judicious Use of Medically Important Antimicrobial Drugs in Food-Producing Animals*. No. 209, April 13. www.fda.gov/downloads/AnimalVeterinary/GuidanceComplianceEnforcement/GuidanceforIndustry/UCM216936.pdf.

FDA. 2013. *Guidance for Industry: New Animal Drugs and New Animal Drug Combination Products Administered in or on Medicated Feed or Drinking Water of Food-Producing Animals: Recommendations for Drug Sponsors for Voluntarily Aligning Product Use Conditions with GFI #209*. No. 213, December. www.fda.gov/downloads/AnimalVeterinary/GuidanceComplianceEnforcement/GuidanceforIndustry/UCM299624.pdf.

FDA. 2015. Veterinary feed directive. *Federal Register* 80(106), 31708–31735.

O'Neill, J. 2014. *Antimicrobial Resistance: Tackling a Crisis for the Health and Wealth of Nations*. UK Welcome Trust. http://amr-review.org/sites/default/files/AMR%20Review%20Paper%20-%20Tackling%20a%20crisis%20for%20the%20health%20and%20wealth%20of%20nations_1.pdf.

Schmithausen, R.M., et al. 2015. Analysis of transmission of MRSA and ESBL-E among pigs and farm personnel. *PLoS One*. DOI:10.1371/journal.pone.0138173.

Sneeringer, S., et al. 2015. *Economics of Antibiotic Use in U.S. Livestock Production*. USDA Economic Research Report #2000, November.

Strom, S. 2015. McDonald's to restrict antibiotic use in chicken. *The New York Times*, International ed., March 6, Finance, p. 17.

Sun, J., et al. 2014. Development of aminoglycoside and â-lactamase resistance among intestinal microbiota of swine treated with lincomycin, chlortetracycline, and amoxicillin. *Frontiers in Microbiology* 5, 580.

Tyson Foods. 2018. *Antibiotic Use.* www.tysonfoods.com/news/viewpoints/antibiotic-use.

US Code of Federal Regulations. 2018. Title 7, Sections 62.000, 62.207.

USDA (United States Department of Agriculture). 2002. United States standards for livestock and meat marketing claims. *Federal Register* 67, 79552–79556.

USDA. 2014. *United States National Residue Program for Meat, Poultry, and Egg Products: 2012 Residue Sample Results.* September.

USDA. 2015. *NE3 Marketing Program.* November 13. www.ams.usda.gov/content/ne3-marketing-program.

White House. 2015. *National Action Plan for Combating Antibiotic-Resistant Bacteria.* Washington, DC.

WHO (World Health Organization). 2017. *WHO Guidelines on Use of Medically Important Antimicrobials in Food-Producing Animals.* Geneva.

10 Controversies with hormones

Key questions to consider

1 Why are consumers concerned about hormones being administered to food animals?
2 Does the US government guarantee that hormone use in food animals is safe?
3 Why do producers use hormones?
4 Are hormones causing ecological damages?

Many consumers are leery about hormones being administered to food animals. Scientists have learned how to manufacture hormones and administer them to animals to enhance growth or augment production. Perhaps the most controversial has been the use of recombinant bovine somatotropin (rBST), a synthetically created version of a hormone responsible for regulating a cow's milk production. In addition, there is concern about the use of steroidal hormones in food animal production. Numerous claims and stories abound how US meat products contain hormones and are unhealthy.

Before looking at this issue, two basic facts need to be highlighted. First, every meat product from a food animal contains hormones. Anyone eating a meat product is accepting the consumption of hormones. Second, each of us consumes non-meat products causing hormonal activity. We cannot avoid hormones. The issue is whether the administration of additional natural and synthetic hormones to food animals from which we harvest food products negatively affects people's health.

Hormones being used

Currently, there are eight approved uses of hormones in food animals in the United States. In the 1970s, scientists discovered how to use recombinant DNA processes to develop rBST. This artificial hormone could be administered to cows to supplement their naturally occurring bovine somatotropin to increase milk production. By allowing each cow to give more milk, rBST means fewer dairy animals are needed to meet the nation's milk needs.

Table 10.1 Hormones used in US beef production*

Hormone	Source	Steroid class	Treatment	Initial FDA approval
Estradiol benzoate	Natural	Estrogenic	Ear implant	1956
Estradiol 17β	Natural	Estrogenic	Ear implant	1991
Melengestrol acetate	Synthetic	Estrogenic	Feed additive	1968
Progesterone	Natural	Estrogenic	Ear implant	1956
Testosterone	Natural	Androgenic	Ear implant	1958
Trenbolone acetate	Synthetic	Androgenic	Ear implant	1987
Zeranol	Non-natural chemical	Estrogenic	Ear implant	1969

* Schweihofer and Buskirk, 2016

The remaining seven hormones are for cattle and sheep (Table 10.1). Hormone treatments have not been approved for pigs or poultry and cannot be administered to veal calves. Dairy cows do not receive these hormones. Given the low consumption of products from sheep by Americans, the main concern is the use of hormones in beef cattle (Centner, 2017). Any consumer expressing concern about hormones in veal, pork, or poultry products is misinformed.

Four of these hormones are natural: they include estradiol benzoate (estrogen), estradiol 17β, progesterone, and testosterone. Three hormones are non-natural hormones: zeranol, trenbolone acetate, and melengestrol acetate. Some of these hormones were approved for use more than 50 years ago.

The seven hormones are either estrogenic or androgenic steroids. Five are estrogenic, which means they are primarily female sex hormones. Two of the hormones are androgenic, which means they stimulate the development of male characteristics by binding to androgen receptors.

Concerned consumers may wonder whether unapproved hormones are being used by producers on other food animal species. This is highly unlikely. For some species, like poultry, it is uneconomical to use hormones. For most producers, the financial incentives are too low to risk their livelihoods. Any breach of the law will result in the termination of a producer's marketing contract, and other buyers will be hesitant to purchase animals from someone identified as unlawfully treating animals with hormones.

Recombinant bovine somatotropin

Bovine somatotropin (BST) appears naturally in all milk cows. By using rBST, producers can increase milk production per unit of food consumed by the cows. This reduces the need for land for growing crops to be fed to dairy cows and amounts of manure disposed on cropland and pastures. The US Food and Drug Administration (FDA) approved the use of rBST in 1993 (FDA, 1993). Monsanto proceeded to market it under the trade name Posilac® in 1994. Posilac® is currently marked by Elanco.

The laudable features accompanying the use of rBST are completely overshadowed by concerns about human and animal health. The major concern is whether milk from cows injected with rBST is safe to drink. A second concern is whether it adversely affects the cows. These health concerns have been accompanied by efforts to require labeling, prohibit information, and boycott products from cows that were treated with rBST.

Before an animal drug may be distributed for commercial applications in the United States, it must be approved by the FDA. The review process considers the safety of the target animal, safety of food from treated animals for human consumption, and environmental safety. To assure the safety of rBST, the FDA analyzed the cumulative effect of continued use on treated animals and on humans consuming the food.

The first issue is safety of products from rBST-treated cows for humans. The FDA found no evidence that rBST posed a health threat to humans. Moreover, the American Medical Association, American Dietetic Association, Congressional Office of Technology Assessment, and National Institutes of Health found no reason to preclude the marketing of products from cows treated with rBST (Mayer et al., 1995).

Although governmental and professional groups in the United States failed to find a legal reason to preclude the use of rBST, it has not enjoyed a similar reception in the rest of the world. In 2000, the European Union (EU) decided to ban the use of rBST because of uncertainty concerning human health implications. Canada, Australia, New Zealand, and Japan have also banned its use.

The second health issue involves the health of cows being treated. The most frequently discussed issue involves evidence that the use of rBST in cattle was accompanied by increases of mastitis. This involves a swelling of the cow's udder characterized by the presence of abnormal milk. However, a majority of the research supports a finding that the use of rBST has no adverse effects on cows' health (Bauman et al., 2015). Moreover, the sale of rBST is accompanied by a warning that users should implement mastitis management practices in their herds.

Researchers also looked at lameness, problems with breeding, life spans of animals, clinical diseases, and retained placentas. While some research identified these issues as problems for treated cows, the only issue that has a significant split in results was the problem of retained placentas.

With the approval of the use of rBST, questions arose about whether milk and milk products from cows receiving rBST should be labeled. State governments and industry requested this guidance from the FDA so that regulations would be the same throughout the nation. Guidance on labeling was issued in 1994 (FDA, 1994). This guidance provided the basis for labeling products and did not prohibit states from adopting additional provisions. Most states allow voluntary labeling with truthful information that does not mislead consumers.

Although a majority of consumers may not be concerned about whether their milk came from supplemented cows, marketing firms do care (Dudlicek, 2009). To augment their public image of selling healthy food products, several supermarket firms do not purchase milk from herds where rBST is used (Dudlicek, 2009). Walmart, Kroger, and Publix supermarkets advertise that their milk products come from cows that were not administered rBST. These three supermarket chains sell more than 40 percent of the groceries purchased in the United States.

Hormones administered to cattle

Producers treat cattle with hormones to increase the animals' growth rate and the efficiency by which they convert the feed they eat into meat. The additional hormone allows animals to use less feed in producing their meat products. This reduces production costs and increases profits.

Hormones are usually administered to cattle by an ear implant. A small pellet is inserted into the back of the ear that releases the hormone over a period of time. Since excessive residual amounts may remain in the ear, it is discarded at harvest so the implant does not enter any food product. An estimated 92 percent of cattle in American feedlots are implanted at least once during their lifetime due to the economic benefits of the procedure (Reuter et al., 2016).

Hormone implants can be used on calves, animals grazing in pastures, and animals being finished for marketing to help speed up their growth. Many implants are only effective for 90 to 120 days, so an animal may receive two or three implants (Reuter et al., 2016).

Producers will consider a number of factors in deciding whether to treat animals. These include the age of the animal, the breed, dosage rates, and whether the animal is expected to be used as breeder stock. Implants should not be used on breeder bulls. Research suggests that implants are a cost-effective technology, returning $15 for every $1 invested (Reuter et al., 2016).

FDA approval of hormones

The use of estrogenic and androgenic hormones in the production of food animals is regulated by the FDA. The FDA examined studies on the safety of the use of each hormone on food animals, the environment, and humans (FDA, 2015). From the scientific data, the FDA established acceptable safe limits for hormones in meat products. The amounts of hormone left in edible tissue after treatment were below appropriate safe levels. The FDA's evaluations included public input and all of the data, studies, and its conclusions are available to the public.

With the use of natural hormones (estradiol, progesterone, and testosterone), people are not at risk from eating food products from treated animals because the amounts of additional remaining hormones are very small. For synthetic hormones, the FDA requires that the amount of the hormone in each edible tissue be below an appropriate safe level. A safe level is a level which would be expected to have no harmful effect in humans (FDA, 2015).

The safety of beef products from animals receiving hormones can be evaluated by the comparisons set forth in Table 10.2. The amounts of estradiol, progesterone, or testosterone that we consume by eating beef from a treated steer are low

Table 10.2 Steroid levels in humans and in 300 grams (10.58 ounces) of beef products from animals treated with hormones measured in micrograms per day*

Hormone	Daily production by humans	Residue in muscle of treated animals	Amount humans receive from a beef product
Estradiol	< 14 (prepubertal boys) 10~24 (prepubertal girls) 27~68 (adult men) 30~470 (adult women)	0.011~0.28	0.0033~0.084
Progesterone	150~250 (prepubertal children) 416~750 (adult men, premenopausal women)	0.230~0.770	0.069~0.231
Testosterone	30~100 (prepubertal children) 210~480 (adult female) 2100~6900 (adult male)	0.031~0.360	0.0093~0.108

* Jeong et al., 2010

compared to the amount we produce ourselves. For example, for testosterone, young children produce at least 30 micrograms per day, while the consumption of 10.58 ounces of beef might have one-tenth of a microgram. The additional amounts of hormones in meat from treated animals are negligible.

Estrogenic comparison

An evaluation of amounts of estrogen in beef as compared to the estrogenic activity of other food products suggests that consumption of beef from treated animals should not have significant effects on estrogen levels in humans. This is because our bodies produce natural estrogen every day that is released into our bodies, and we consume estrogen with other food products. Table 10.3 shows the estrogenic activity of several food products and the production of estrogen by humans.

For example, a slice of white bread may lead to 30 micrograms of estrogenic activity, whereas a 3-ounce serving of beef from an implanted animal would contain 0.002 micrograms (North American Meat Institute, 2016). Moreover, three ounces of beef from a treated animal as compared to a non-treated animal would have less than one thousandth of a microgram more estrogen. These data suggest that eating beef from treated cattle does not markedly alter our estrogen levels.

Human concerns

Despite the government's safety assurances, some people fear that hormones administered to animals will remain in food products or the environment, causing increases in the risk of cancer or overall human health problems. While some studies have led to conclusions that the consumption of meat products from animals administered hormones may be related to human health issues, most of

Table 10.3 Estrogenic activity in micrograms per 17.64 ounces of common foods and the daily production of estrogen by humans*

Food product[†]	Estrogenic activity	Production in humans[†]	Micrograms/day
Tofu	113,500	Pregnant woman	19,600
Pinto beans	900	Non-pregnant woman	513
White bread	300	Adult man	136
Peanuts	100	Pre-pubertal children	41
Milk	0.032		
Beef: implanted steer	0.007		
Beef: non-implanted steer	0.005		

* Loy, 2011
† Hoffman and Eversol, 1986; Hartman et al., 1998; Shore and Shemesh, 2003; USDA, 2002. Units are micrograms of estrone plus estradiol for animal products and isoflavones for plant products

the claims and stories of problems from hormones cannot be substantiated by reputable scientific studies. This is because it is very difficult to separate added hormones in meat products from the hormones produced by the person or those that come from other food items.

Scientific studies involving the consumption of food are also challenging because other foods contain hormonal materials. Can the study distinguish added hormones in meat from hormones in other foods and natural hormones? When an animal receives an additional natural hormone, such as testosterone, it becomes indistinguishable from the animal's natural testosterone. This means that identifying supplemental natural hormones as creating a problem is nearly impossible.

Exacting human studies on the effects to eating meat from treated animals cannot be conducted for ethical reasons. As the anatomy and physiology of pigs are similar to humans, a study on pigs to determine the effects of animals treated with hormones can serve as a substitute. A team of researchers conducted a detailed study used young female pigs to determine whether the consumption of cooked beef patties from steers treated with trenbolone acetate and estradiol benzoate had any effect on their puberty (Magolsky et al., 2014). The study concluded that the consumption of meat from treated animals did not affect puberty of the young pigs.

Oftentimes research studies involve feeding large quantities of hormones to other animal species, such as rodents. These studies may not provide an accurate accounting of the effects of the hormones on humans. Given the difficulties of separating added hormones from treating food animals with other sources of hormones, no conclusive results showing adverse human health effects exist (Biswas et al., 2013). Other criteria such as diets, sexual activity, reproductive behavior, and genetic susceptibility probably account for any claimed relationships between ingesting meat from treated animals with human health problems.

Another concern is that drinking water containing steroids from hormone use in cattle may expose humans to excessive amounts of steroids. However, our water sources have very low levels of natural or synthetic steroids (Biswas et al., 2013).

Ecological concerns

We do have evidence that accumulations of steroids in soils or leachates from areas where animals are housed or lands where their manure and urine are deposited are causing problems. Hormones from animals are carried in overland runoff to surface waters. The concentrations of hormones may be at levels that negatively impact the health of aquatic organisms (Bartelt-Hunt et al., 2012). A major problem is that the concentrations of hormones cause an endocrine disruption in fish (Orlando et al., 2004).

The problem is the presence of too many hormones in soil or water environments. However, while treating food animals with hormones contributes to the problems, it is not its sole source. All vertebrates excrete hormones. Non-treated animals and humans also contribute hormones to the environment (Bartelt-Hunt et al., 2012). Whenever too many hormones are added to an area, they can have adverse effects on the ecology of an area, especially aquatic species. Additional hormones can alter the sexual differentiation of fish (Thornton et al., 2016).

Studies on waste from species like pigs and chickens that are not treated with hormones show hormones entering water sources. Steroids from human usage entering wastewater treatment plants are polluting water supplies. Research studies on soil and water contamination from animals often note that the observed concentrations of hormones cannot be proven to have come from food animals treated with hormones.

While we need to worry about too many hormones, low detection rates of steroid hormones in soil and leachate samples suggest that most hormones from food animals are readily degraded or adsorbed after manure application (van Donk et al., 2013). The storage of manure in lagoons can degrade 99 percent of the estrogen potentially excreted by cows (Zhao et al., 2010).

Simultaneously, we do not know enough about the effects of low levels of steroids on human health, so some caution is warranted. Additional research is needed to learn whether the use of hormones in beef production is causing ecological damages.

The EU's ban of hormones

The European Union has prohibited the use of added hormones in animal production since 1981. In 1989, the EU prohibited imports of meat products from animals that have been administered hormones, with the justification that it was necessary to protect consumer health and safety (Johnson, 2015). The EU's scientific committee reported that no study has assessed the effects of hormones as growth promoters in farm animals on cancer occurrence in humans (European

Commission, 1999). In the absence of research showing usage of hormones to be safe, the Commission concluded that scientific information was insufficient to approve the use of six hormones under consideration.

The EU's ban on meat imports from animals administered hormones led the United States to initiate a trade dispute in the World Trade Organization. After a series of deliberations and rulings, the EU and US entered a Memorandum of Understanding in 2009 delineating a quota for imports of non-hormone treated beef. In 2018, the EU announced that it would discuss with the US the functioning of its existing quota to resolve this disagreement in line with World Trade Organization rules.

The European public continues to express its support for banning hormone use in food animal production. They view hormone use in food animals very differently than most Americans. The ban precludes imports of meat products from treated animals but allows imports of products from non-treated cattle into the EU.

Several countries have expressed an unwillingness to import beef products from animals receiving hormones. Since producers in Argentina, Brazil, Canada, Australia, and New Zealand also allow hormone use in beef cattle, they are also confronted with the choice of whether to use hormones. Producers in all countries have the option of forgoing hormone use so their animals produce food items that can be sold in these foreign markets.

Misguided research

In 2016, two university researchers reported a study involving factors that influence consumers' preferences for chicken products (Samant and Seo, 2016). The researchers wanted to see if chicken products labeled with different production practices had an effect on how people viewed the product's tenderness, juiciness, and chicken flavor. Their chicken products were labeled "No Hormones Added," "USDA Organic," "USDA Process Verified," and "no-label." The test participants tasted the same chicken meat from the four differently labeled samples and then rated the products.

What the authors neglected to tell participants or persons reading about their research is that US regulations prohibit the use of hormones in raising poultry (USDA, 2011). A claim of "No Hormones Added" cannot be used on the labels of poultry unless it is followed by a statement that says "Federal regulations prohibit the use of hormones" (USDA, 2011). The authors did not use this statement (Samant and Seo, 2016).

This study was flawed because the assumption that a chicken product might contain hormones is false. The label used by the researchers was unlawful. No consumer would have ever seen such a label. These facts raise the question whether some participants knew about the falsehood or how they would react to a label they had never seen. Another question is why the journal would publish research based on an unlawful label.

Equally as troubling is that this research presented misinformation about American poultry production. Many readers would interpret the study to infer that the US Department of Agriculture allows the use of hormones in the production of poultry. This is not true and unfairly raises issues about the competence of the American government and the integrity of the US poultry industry.

Finally, by reporting research on an unlawful label, this study exacerbates consumer confusion. Accurate, truthful information about which animal species are receiving hormone implants or infusions is needed. Given the invented use of an unlawful label in the authors' study, it seriously detracts from consumers' trust in label claims.

Foodwashing facts

1   Meat products from pigs and poultry do not contain any added hormones.
2   The residual hormones in meats from treated animals are low compared to hormones produced by the human body.
3   Many cattle in the US are implanted with a hormone pellet to increase their rate of growth.
4   By enhancing animal growth, less feed and fewer animals are needed to supply desired beef products.

References

Bartelt-Hunt, S., et al. 2012. Effect of growth promotants on the occurrence of endogenous and synthetic steroid hormones on feedlot soils and in runoff from beef cattle feeding operations. *Environmental Science and Technology* 46, 1352–1360.

Bauman, D.E., et al. 2015. An *Updated Meta-Analysis of Bovine Somatotropin: Effects on Health and Welfare of Dairy Cows*. 24th Tri-State Dairy Nutrition Conference, Fort Wayne, Indiana.

Biswas, S., et al. 2013. Current knowledge on the environmental fate, potential impact, and management of growth-promoting steroids used in the US beef cattle industry. *Journal of Soil and Water Conservation* 68(4), 325–336.

Centner, T.J. 2017. Differentiating animal products based on production technologies and preventing fraud. *Drake Journal of Agricultural Law* 22(2), 267–291.

Dudlicek, J. 2009. BST party. *Dairy Foods* 110(5), 8.

European Commission. 1999. *Opinion of the Scientific Committee on Veterinary Measures Relating to Public Health: Assessment of Potential Risks to Human Health from Hormone Residues in Bovine Meat and Meat Products*. Brussels: Directorate General XXIV, Scientific Health Opinions.

FDA (Food and Drug Administration). 1993. Animal drugs, feeds, and related products: Sterile sometribove zinc suspension. *Federal Register* 58, 59946–59947.

FDA. 1994. Interim guidance on the voluntary labeling of milk and milk products from cows that have not been treated with recombinant bovine somatotropin. *Federal Register* 59, 6279–6280.

FDA. 2015. *Steroid Hormone Implants Used for Growth in Food-Producing Animals*. www.fda.gov/AnimalVeterinary/SafetyHealth/ProductSafetyInformation/ucm055436.htm.

Hartmann, S., et al. 1998. Natural occurrence of steroid hormones in food. *Food Chemistry* 62, 7–20.

Hoffman, B., Eversol, P. 1986. Anabolic agents with sex hormone-like activities: Problems of residues. In *Drug Residues in Animals*, A. G. Rico (ed.), pp. 111–146. New York: Academic Press.

Jeong, S-H., et al. 2010. Risk assessment of growth hormones and antimicrobial residues in meat. *Toxicological Research* 26(4), 301–313.

Johnson, R. 2015. The U.S.-EU beef hormone dispute. *Congressional Research Service* 7-5700.

Loy, D. 2011. *Understanding Hormone Use in Beef Cattle.* Iowa Beef Center, Iowa State University. www.iowabeefcenter.org/information/IBC48.pdf.

Magolsky, J.D., et al. 2014. Consumption of ground beef obtained from cattle that had received steroidal growth promotants does not trigger early onset of estrus in prepubertal pigs. *Journal of Nutrition* 144, 1718–1724.

Mayer, R.N., et al. 1995. Consumer risk perception and recombinant bovine growth hormone: The case for labeling dairy products made from untreated herd milk. *Journal of Public Policy & Marketing* 14(2), 328–300.

North American Meat Institute. 2016. *Growth Promotants in Meat Production: Their Use and Safety.* Washington, DC. www.meatinstitute.org/index.php?ht=a/GetDocumentAction/i/125340.

Orlando, E.F., et al. 2004. Endocrine-disrupting effects of cattle feedlot effluent on an aquatic sentinel species, the fathead minnow. *Environmental Health Perspectives* 112(3), 353–358.

Reuter, R., et al. 2016. *Implants and Their Use in Beef Cattle Production.* ANSI 3290. Oklahoma Cooperative Extension Service. http://pods.dasnr.okstate.edu/docushare/dsweb/Get/Document-10027/ANSI-3290web.pdf.

Samant, S.S., Seo, H-S. 2016. Quality perception and acceptability of chicken breast meat labeled with sustainability claims vary as a function of consumers' label-understanding level. *Food Quality and Preferences* 49, 151–160.

Schweihofer, J.P., Buskirk, D.D. 2016. *Farm to Fork – Part 1 – Antibiotics and Hormones.* East Lansing, MI: Michigan State University Department of Animal Science. http://fec.msue.msu.edu/uploads/files/2016_FEC/Hormones_-_Buskirk_-_FEC.pdf.

Shore, L.S., Shemesh, M. 2003. Naturally produced steroid hormones and their release into the environment. *Pure Applied Chemistry* 75(11–12), 1859–1871.

Thornton, L.M., et al. 2016. Early life stage exposure to BDE-47 causes adverse effects on reproductive success and sexual differentiation in fathead minnows (Pimephales promelas). *Environmental Science and Technology* 50, 7834–7841.

USDA (US Department of Agriculture). 2002. *USDA-Iowa State University Database on the Isoflavone Content of Foods, Release 1.3–2002.* Agricultural Research Service.

USDA. 2011. *Meat and Poultry Labeling Terms.* Food Safety and Inspection Service. www.fsis.usda.gov/wps/wcm/connect/e2853601-3edb-45d3-90dc-1bef17b7f277/Meat_and_Poultry_Labeling_Terms.pdf?MOD=AJPERES.

van Donk, S.J., et al. 2013. Transport of steroid hormones in the vadose zone after land application of beef cattle manure. *Transactions of the American Society of Agricultural and Biological Engineers* 56(4), 1327–1338.

Zhao, S., et al. 2010. Estrogens in streams associated with a concentrated animal feeding operation in upstate New York, USA. *Chemosphere* 79, 420–425.

11 Health concerns with beta agonist feed additives

Key questions to consider

1 What are beta agonists?
2 Why are low amounts of residues from beta agonists allowed in meat products?
3 Why did a United Nations Committee approve residues of ractopamine in food products?
4 Why does the European Union not allow beta agonists to be used in food animal production?

Another production tool used by producers of food animals is feed additives. These are supplements in feed for livestock that include vitamins, amino acids, fatty acids, and minerals. Additives are used to enhance animals' appetites, to provide deficient nutrients, to improve the nutritional value of diets, and to inhibit protozoa growth. For food animals, feed additives are used for all of these purposes.

While we want to have healthy animals and this may include some uses of feed additives, the use of two beta agonist drugs has become controversial. The drugs are ractopamine and zilpaterol hydrochloride. Ractopamine was approved by the US Food and Drug Administration (FDA) for use in swine in 2000 and cattle in 2003, and zilpaterol was approved by the FDA for use in cattle in 2006.

Adding beta agonists to animals' feed

Beta agonists are organic molecules that activate protein synthesis and decrease protein degradation on a cellular level. Ractopamine and zilpaterol enhance animals' muscle growth and limit the amount of fat in meat products without increasing natural hormone levels (Gonzalez et al., 2007). These beta agonists also lead to greater feed efficiency, protein accretion, and improved meat yield grade. These features mean that the use of beta agonists can be profitable for producers (López-Campus et al., 2013). Ractopamine is used in more than 20 countries, while zilpaterol has been used by livestock producers in five countries.

For cattle and swine, a number of beta agonist feed additive products exist (Table 11.1). The first is Optaflexx®, manufactured by Elanco. Optaflexx®

Table 11.1 Beta agonists being fed to food animals

Beta agonist	Market name	Manufacturer	Animal species
Ractopamine hydroxide	Optaflexx®	Elanco	Cattle
Ractopamine hydroxide	Paylean®	Elanco	Swine and heavy turkeys
Ractopamine hydroxide	Engain™	Zoetis Inc.	Swine
Ractopamine hydroxide	Actogain™	Zoetis Inc.	Cattle
Zilpaterol hydroxide	Zilmax®	Merck Animal Health	Cattle
Ractopamine hydroxide	Ractopamine 100	Bio Agri Mix	Cattle

contains the active ingredient ractopamine hyperchloride and is used for cattle. A second is Paylean®, which is also manufactured by Elanco and contains the active ingredient ractopamine hydrochloride; it is used for swine and heavy turkeys. A third is Zilmax®, manufactured by Merck Animal Health, which contains the active ingredient zilpaterol hydrochloride and is used for cattle. Zoetis Inc. manufactures Engain™ and Actogain™ for swine and cattle. Bio Agri Mix, a Canadian firm, manufactures Ractopamine 100 as a feed additive for cattle.

These veterinary drugs need to be used according to specific dosing regimens. Their use is generally limited to a few weeks immediately prior to slaughter. For cattle, ractopamine may be used 28–42 days prior to slaughter (FDA, 2003). Zilpaterol is fed to cattle the last 20–40 days. Ractopamine is fed to hogs weighing at least 150 pounds (68.04 kg) prior to slaughter. Use of these additives leaves residues in the meat products.

American and Canadian livestock producers feed their animals beta agonists because of the economic advantages. By improving feed efficiency and stimulating muscle growth, producers using beta agonists have lower costs per hundredweight of salable animal products. This competitive advantage has been important in facilitating the sale of beef and swine meat products all over the world.

Yet, most countries do not allow the sale of meat products containing residues of ractopamine or zilpaterol. Their decisions are based on concerns about potential human health problems (Bories et al., 2009). Concern also exists about the welfare of animals (Loneragan et al., 2014). The major pork producers are discontinuing the use of ractopamine in pig production, and many of their processing facilities only process animals that did not receive feed containing ractopamine.

Beta agonist residues and their health effects

Because residues of beta agonists remain in meat products, the use of beta agonists had to be approved by the FDA to be sure that the food products were safe for human consumption. The FDA calculated the amount of total residue that a human can safely consume per day over a lifetime (FDA, 2012). This involved the calculation of acceptable daily intakes and the calculation of tolerance

Table 11.2 FDA tolerance levels in parts per million and acceptable daily intakes in micrograms per kilogram of body weight (ug/kg bw)*

Beta agonist	Species	Tolerance level	Acceptable daily intake
Ractopamine	Cattle	0.09 ppm for liver	1.25 µg/kg bw/day
		0.03 ppm for muscle	1.25 µg/kg bw/day
	Swine	0.15 ppm for liver	1.25 µg/kg bw/day
		0.05 ppm for muscle	1.25 µg/kg bw/day
	Turkeys	0.45 ppm for liver	1.25 µg/kg bw/day
		0.10 ppm for muscle	1.25 µg/kg bw/day
Zilpaterol	Cattle	0.012 ppm for liver	0.083 µg/kg bw/day
		0.010 ppm for muscle	0.083 µg/kg bw/day

* US Code of Federal Regulations, 2018

levels for each edible meat tissue so that a person's consumption would be safe (Table 11.2).

A tolerance level for beta agonists is the maximum concentration of a residue that can legally remain in a specific edible tissue of an animal product. Since cattle and swine are different species, separate limits were calculated for ractopamine residues. Furthermore, different residue limits were calculated for edible liver as opposed to muscle tissues (US Code of Federal Regulations, 2018).

For humans, the FDA adopted an acceptable daily intake for ractopamine hydrochloride of 1.25 µg/kg bw (micrograms per kilogram of body weight) per day. For zilpaterol, the acceptable daily intake for total residues is 0.083 µg/kg bw. With the approval of these intakes, meat products can be sold in the United States containing residues of beta agonists.

Some animals that had been fed the beta agonist zilpaterol developed ambulatory problems (Thomson et al., 2014). The cattle seemed to have sloughed hoof walls, resulting in extreme discomfort that adversely affected their welfare. Given the lameness of cattle and concerns about the animals' welfare, two large beef packers declined to accept animals that had been fed zilpaterol.

The drug maker temporarily pulled its zilpaterol product from the market, and research studies were conducted to discern whether the problem with the lame cattle was related to the use of zilpaterol. Other factors may have contributed to the problem. However, some experts concluded the cattle had fatigued cattle syndrome.

In addition, the European Food Safety Authority looked at research studies investigating the use of zilpaterol. Although the research suggested that cattle fed zilpaterol had an increase in mortality, heart rate, respiration rate, and agonistic behavior, there was insufficient evidence to conclude that the observed effects were related to the use of zilpaterol (Arcella et al., 2016).

United Nations health standards

The world recognizes the United Nations' Food and Agriculture Organization (FAO) and the World Health Organization (WHO) as institutions that develop guidance for protecting foodstuffs and human health. Since 1956, these two organizations have worked together by forming a combined Joint FAO/WHO Expert Committee on Food Additives (UN Committee) to provide guidance on the use of food additives. The UN Committee consists of international scientific experts who focus on evaluating contaminants, naturally occurring toxicants, and residues of veterinary drugs in food (FAO, 2013a).

Working in conjunction with the UN Committee, the Codex Alimentarius Commission "develops harmonized international food standards, guidelines and codes of practice to protect the health of the consumers and ensure fair practices of the food trade" (Codex Alimentarius Commission, 2012). The Codex standards enumerate guidance for nations to use in developing national legislative provisions to provide consumers safe and wholesome food products (FAO, 2010a).

The FAO and WHO have complementary functions in selecting experts to serve on the UN Committee (WHO, 2013). With respect to beta agonist residues, the UN Committee evaluated residues of veterinary drugs to recommend acceptable daily intakes and tolerance levels. The UN's tolerance levels are known as maximum residue limits (MRLs).

An MRL is the maximum concentration of residue resulting from the use of a veterinary drug (expressed in mg/kg or ug/kg on a fresh weight basis) permitted or recognized as acceptable in or on a food. An MRL is based on the type and amount of residue considered to be without any toxicological hazard for human health as expressed by the acceptable daily intake or on the basis of a temporary daily intake that utilizes an additional safety factor. The MRL also takes into account other relative public health risks as well as food technological aspects.

In 2004, the UN Committee established an acceptable daily intake for ractopamine and presented recommendations to the Codex Alimentarius Commission's Committee on Residues of Veterinary Drugs in Foods. The UN Committee also recommended MRLs for residues of ractopamine in meat tissues.

In 2010, the UN Committee reviewed new data of ractopamine residues in pig lung, heart, and intestinal tissues, taking into account studies and data provided by the People's Republic of China (FAO, 2010b). China was concerned that pig organs are consumed as part of a traditional Chinese diet so that the delineations of maximum residue levels might be insufficient.

The UN Committee considered the submitted studies from China and concluded that the recommended ractopamine MRLs for muscle, liver, kidneys, and fat were still below the upper bound of the acceptable daily intakes (FAO, 2006, 2010b). The data on organ tissues from the heart, lung, stomach, and intestine were inconclusive, so more studies were recommended (FAO, 2010b). The UN Committee's recommendations are presented in Table 11.3.

In 2012, the UN Committee established an acceptable daily intake for zilpaterol hydrochloride (FAO, 2013b). However, the Committee concluded that there were inadequate data to establish MRLs for zilpaterol. The Committee listed three needs for data: (1) results from studies investigating marker residue in liver and kidney; (2) results from studies determining marker residue to total

Table 11.3 Joint FAO/WHO Expert Committee on Food Additives tolerance levels and acceptable daily intakes in micrograms per kilogram of a person's body weight*

Beta agonist	Species	Tolerance level	Acceptable daily intake
Ractopamine	Cattle and swine	10 µg/kg for muscle and fat	Less than 1.00 µg/kg
		90 µg/kg for kidney	Less than 1.00 µg/kg
		40 µg/kg for liver	Less than 1.00 µg/kg
Zilpaterol	Cattle	None approved	Up to 0.04 µg/kg

* FAO and WHO, 2006; WHO and FAO, 2015

residue ratio in liver and kidney; and (3) results from depletion studies to enable the derivation of MRLs compatible with the acceptable daily intakes.

A comparison of data from Tables 11.2 and 11.3 shows that the UN Committee recommended more stringent acceptable daily limits than established by the United States (FAO and WHO, 2006; WHO, 2004a).

International Codex standards

The Codex Alimentarius Commission discussed ractopamine in a number of sessions without adopting the MRLs recommended by the UN Committee (Codex Alimentarius Commission, 2012). In 2012, however, the Codex Alimentarius Commission adopted MRLs for ractopamine in cattle and pig tissues (Codex Alimentarius Commission, 2012). The process of adopting the Codex MRLs occurred through a series of secret ballot votes and resulted in a 69–67 majority in favor of adopting the MRLs.

While the United States was pleased with the majority vote adoption of MRLs, the European Union, Norway, China, and eight other members of the Commission expressed their concerns (Codex Alimentarius Commission, 2012). Several countries in opposition felt that possible risks to human health existed. Another objection by some opponents was the fact that decisions on international standards should require a consensus rather than majority vote. Specifically, the delegation from Norway noted that "the draft MRLs had been pushed forward when many members had asked for a consensus-based decision and the MRLs had been adopted despite a clear lack of consensus" (Codex Alimentarius Commission, 2012).

The United States noted that the adoption of the standard should be "a rare exception to the general principle of consensus" (Codex Alimentarius Commission, 2012). The lack of consensus on ractopamine MRLs and human health concerns may lead to differences in opinion about the meaning of the Codex standards under the World Trade Organization Agreement on the Application of Sanitary and Phytosanitary Measures.

The significance of the Codex residue limits is that they need to be followed by countries that are members of the World Trade Organization to comply with the obligations of the Agreement on Technical Barriers to Trade and the Agreement on the Application of Sanitary and Phytosanitary Measures. These member countries have agreed not to impose more strict standards than adopted by Codex that might interfere with free trade.

European Food Safety Authority

For the European Union, the European Food Safety Authority (EFSA) conducts safety evaluations of veterinary drug residues that may be permitted in food. In 1996, three years after the first UN Committee evaluation of ractopamine but well before ractopamine came to market as a livestock growth promoter, the Commission of the European Communities imposed a general ban on the use of beta agonists with farm animals (EU Council Directive, 1996).

The EU had not conducted any studies on ractopamine before adding it to the list of banned veterinary drugs. In response to the UN Committee's 2006 reconfirmation of an acceptable daily intake and MRLs for ractopamine, a panel of the EFSA conducted a safety evaluation of ractopamine in 2009 (Bories et al., 2009). While the examination did not introduce any new research, the EFSA panel took into account all available information about ractopamine, including studies on pigs, cattle, laboratory animals, dogs, monkeys, and humans (Bories et al., 2009).

Because the data from studies of ractopamine in laboratory animals gave a large range of results, the EFSA panel found that human data were of primary concern. Both the EFSA panel and the UN Committee assessed consumer safety for the development of an acceptable daily intake and MRLs for ractopamine by examining the results of one human study (Bories et al., 2009). The study looked at indices of cardiovascular function and safety to increasing doses of ractopamine (WHO, 2004b). Researchers gave six healthy male volunteers placebo and ractopamine beginning at 5 milligrams and increased the dose to 40 milligrams over the course of five doses (WHO, 2004b, p. 148). Data on 14 cardiovascular variables were obtained. While no serious adverse effects were reported, heart rates were elevated with the three higher doses.

While the human study proved sufficient for the UN Committee in determining an acceptable daily intake and MRLs for ractopamine use in livestock, the EFSA panel expressed concern about methods used in the experiment. In particular, the EFSA report found that six subjects did not provide a sufficient sample size for the responses to ractopamine to be statistically significant (Bories et al., 2009). Moreover, one man was withdrawn from the study due to adverse cardiac effects (WHO, 2004b). In conclusion, the EFSA report found that a number of weaknesses and uncertainties limited meaningful conclusions from the study.

Thus, the EFSA panel decided that no MRLs could be established because no conclusion could be rendered on the safety of ractopamine residues in meat products consumed by humans (Bories et al., 2009). In the absence of a conclusion that the consumption of ractopamine residues by humans was safe, the detailed scientific investigation did not provide support to overturn the earlier decision by the Commission of the European Communities banning ractopamine.

Assessing risk

Risk plays a crucial role in the regulation of foods and is especially significant since scientific evidence is often inconclusive. Although scientific risk assessments may be perceived to be value-free, neutral, and unbiased, scientific inquiry incorporates notions of risk and conceptions of scientific evidence. Technologies include uncertainties, choices for perspectives, and value-loadings, so are not comprised of stand-alone, factual answers. Rather, the evaluation of risk involves defining risks, analyzing inconclusive scientific evidence, making assumptions, assigning probabilities, and discerning acceptable levels of potential damages.

Whenever scientists, legislators, and regulators define risk for decisionmaking purposes related to food, their analyses bring logic, reason, and scientific deliberation to bear on the management of food safety. They determine how the risks are framed. Assessments are embedded with non-science values due to the risk assessors' behavior, assumptions, selected processes, and value judgments, which lead to different estimates of scientific risk.

Moreover, a growing body of evidence suggests that experts can have their own biases and agendas. Biases that impact scientific risk assessment can range from overconfidence to conflicts of interest, particular ideologies, and "wish" bias (Weed, 2007). Value judgments and bias in science can result in unwarranted variations in the development of scientific experiments and the interpretation of scientific evidence. Thus, risk assessments are not neutral exercises of scientific evidence.

Although the evaluation of risk accompanying beta agonists involves scientific evidence, legislative and regulatory decisions to protect people, animals, and the environment involve values of the decisionmakers. Decisions may include ethical values In addressing beta agonists, the FDA and the EFSA had nearly identical scientific data about the technologies, yet the agencies' assessments led to different policies. While the United States decided to allow producers to use beta agonists, the EU proscribed beta agonists.

One explanation for the divergent assessment of risk is the heuristics affecting the decisionmakers. The affect and availability heuristics of the decisionmakers who evaluated risks and decided how much precaution was needed to protect people, animals, and the environment may account for the discrepancies in the US and EU regulations on beta agonists.

Applying precaution

It is argued that the divergent transatlantic views toward beta agonists are related to the embracement of the precautionary principle by the EU. In applying the precautionary principle, the EU grants regulators discretion in regulating risk that can lead to products being banned (Morag-Levine, 2014).

However, the US also applies precaution in regulating risk. An examination of a wider range of risks managed by regulatory provisions in the US and EU found that amounts of precaution were similar (Weiner et al., 2014). The study concluded that political and institutional factors, legal systems, and the role of cost–benefit analyses did not account for the observed pattern of precaution applied to risks by the US and EU regulatory provisions. Rather, the authors concluded that "heuristic availability" provided more important clues about the use of the precautionary principle and cross-cultural differences in risk perception (Weiner et al., 2014).

Heuristic availability is access to information that will affect a person's memory (Geurten et al., 2015). By drawing on personal experiences and individual senses, the availability heuristic affects people's perceptions of risk and regulatory oversight (Mase et al., 2015). For beta agonists, persons who evaluated the risks, applied precaution, and authored the regulations had experiences and information that may have impacted their perceptions of the regulatory needs of the technology. Since the technology dealt with food animals, decisionmakers may have been influenced by events and existing regulatory controls over agricultural technologies related to animal well-being and wholesome food.

For decisionmakers in the EU, the affect heuristic may include the memories of deaths from Creutzfeldt–Jakob disease, a human variant of bovine spongiform encephalopathy. Animals can become infected with this disease by eating contaminated feed, a practice that was occurring in the United Kingdom in the 1980s (Budka, 2011). The United Kingdom Creutzfeldt–Jakob Disease Research and Surveillance Unit reported 176 human deaths in the UK from the disease (Andrews, 2012), while only four deaths in total were confirmed for being related to the disease in the US (Centers for Disease Control and Prevention, 2014). Given the seriousness of the outbreak in the EU, decisionmakers in the EU were more likely to have an enhanced concern about the health dangers of contaminated animal feedstuffs.

Legislators and regulators in the EU have a history of being demanding of the safety of new technologies used in animal production. To avoid risks, the EU has embraced the precautionary principle. This has led the EU to adopt regulations forbidding the use of hormones, recombinant bovine somatotropin, and some nontherapeutic antibiotics (Centner and Petetin, 2018). Countries in the European Union also have regulatory controls over concentrations of animals, their confinement cages and crates, and the castration of male animals without anesthesia. Banning products from animals fed beta agonists complements these regulations.

Foodwashing facts

1   Beta agonists are fed to food animals to help them gain weight.
2   Many American beef and pork products contain residues of beta agonists.
3   The FDA has established acceptable daily limits and tolerances for beta agonists so that American meat products are safe for human consumption.
4   Countries in the European Union have adopted a more cautious approach and do not allow beta agonist residues in meat products.

References

Andrews, N.J. 2012. *Incidence of Variant Creutzfeldt-Jakob Disease Diagnoses and Deaths in the UK January 1994 – December 2011*. United Kingdom Creutzfeldt-Jakob Disease Research and Surveillance Unit, July 2.

Arcella, D., et al. 2016. Review of proposed MRLs, safety evaluation of products obtained from animals treated with zilpaterol and evaluation of the effects of zilpaterol on animal health and welfare. *European Food Safety Authority Journal* 14(9), e4579.

Bories, G., et al. 2009. Safety evaluation of ractopamine: ESFA panel on additives and products or substances used in animal feed (FEEDAP). *European Food Safety Authority Journal* 1041, 1–52.

Budka, H. 2011. Editorial: The European response to BSE: A success story. *European Food Safety Authority Journal* 9(9), e991.

Centers for Disease Control and Prevention. 2014. *Confirmed Variant Creutzfeldt-Jakob Disease (variant CJD) Case in Texas*. Atlanta, GA, October 7. www.cdc.gov/ncidod/dvrd/vcjd/other/confirmed-case-in-texas.htm.

Centner, T.J., Petetin, L. 2018. Divergent approaches regulating beta agonists and cloning of food animals: United States and European Union. *Society & Animals* 26(5), 1–20.

Codex Alimentarius Commission. 2012. *Joint FAO/WHO Standards Programme Codex Alimentarius Commission*, Thirty-fifth Session FAO Headquarters. Rome, Italy, July 2–7. Report. REP12/CAC, Rome.

Codex Alimentarius Commission. 2014. *Maximum Residue Limits (MRLs) and Risk Management Recommendations (RMRs) for Residues of Veterinary Drugs in Foods*. CAC/MRL 2–2014. Rome: FAO/WHO.

EU (European Union) Council Directive 96/22/EC of 29 April 1996. 1996. Official Journal of the European Union, No. L 125/3 to 8.

FAO (Food and Agriculture Organization). 2006. *Joint FAO/WHO Expert Committee on Food Additives*, Sixty-sixth meeting (Residues of veterinary drugs). Rome, Italy, pp. 1–18.

FAO. 2010a. *Codex Alimentarius Commission Procedural Manual*. 19th ed. Rome, Italy, p. 1–183.

FAO. 2010b. *Residue Evaluation of Certain Veterinary Drugs*. Joint FAO/WHO Expert Committee on Food Additives Meeting 2010 – Evaluation of data on ractopamine residues in pig tissues. Rome, Italy, pp. 1–51.

FAO. 2013a. *Joint FAO/WHO Expert Committee on Food Additives (JECFA), FAO Roster of experts for JECFA (2012–2016)*. Rome, Italy, pp. 1–7. www.fao.org/publications/search/en/?cx=018170620143701104933%3Aqq82jsfba7w&q=FAO+Roster+of+experts+for+JECFA+%282012-2016%29+&cof=FORID%3A9, March 10, 2014.

FAO. 2013b. *Joint FAO/WHO Expert Committee on Food Additives Seventy-Eighth Meeting (Residues of veterinary drugs) Geneva*, November 5–14, Summary and Conclusions, November 22.

FAO and WHO (World Health Organization). 2006. *Joint FAO/WHO Expert Committee on Food Additives*, Sixty-Sixth Meeting (Residues of veterinary drugs), Rome, February 22–28, pp. 1–18.

FDA (Food and Drug Administration). 2003. New animal drugs: Ractopamine. *Federal Register* 68(181), 54658–54660.

FDA. 2012. New animal drugs: Updating tolerances for residues of new animal drugs in food; proposed rule. *Federal Register* 77(234), 72254–72268.

Gonzalez, J.M., et al. 2007. Effect of ractopamine-hydrochloride and trenbolone acetate on longissimus muscle fiber area, diameter, and satellite cell numbers in cull beef cows. *Journal of Animal Science* 85, 1893–1901.

Geurten, M., et al. 2015. Less is more: The availability heuristic in early childhood. *British Journal of Developmental Psychology* 33, 405–410.

Huffstutter, P.J., Baertlein, L. 2013. For Merck, bringing cattle feed Zilmax back won't be easy. *Reuters*. October 31.

Loneragan, G.H., et al. 2014. Increased mortality in groups of cattle administered the β-adrenergic agonists ractopamine hydrochloride and zilpaterol hydrochloride. *PLOS One* 9(3), 1–13.

López-Campus, Ó., et al. 2013. Effects of calf- and yearling-fed beef production systems and growth promotants on production and profitability. *Canadian Journal of Animal Science* 93, 171–184.

Mase, A.S., et al. 2015. Enhancing the social amplification of risk framework (SARF) by exploring trust, the availability heuristic, and agricultural advisors' belief in climate change. *Journal of Environmental Psychology* 41, 166–176.

Morag-Levine, N. 2014. The history of precaution. *American Journal of Comparative Law* 62, 1095–1131.

Thomson, D.U., et al. 2014. Description of a novel fatigue syndrome of finished feedlot cattle following transportation. *Journal of the American Veterinary Medical Association* 247(1), 66–72.

US Code of Federal Regulations. 2018. Title 21, Sections 556.570, 556.765.

WHO (World Health Organization). 2004a. *Evaluation of Certain Veterinary Drug Residues in Food*. WHO Technical Report Series 925. Geneva, pp. 1–72.

WHO. 2004b. *Toxicological Evaluation of Certain Veterinary Drug Residues in Food*. WHO Food Additives Series 53, Ractopamine addendum. Geneva, pp. 119–164.

WHO. 2013. *About the Joint FAO/WHO Expert Committee on Food Additives (JECFA)*. International Programme on Chemical Safety. Geneva.

WHO and FAO. 2015. *Evaluation of Certain Veterinary Drug Residues in Food*. WHO Technical Report Series 988, Seventy-Eighth Report of the Joint FAO/WHO Expert Committee on Food Additives, Geneva.

Weed, D.L. 2007. The nature and necessity of scientific judgment. *Journal of Law & Policy* 15, 135–164.

Weiner, J., et al. 2014. *The Reality of Precaution: Comparing Risk Regulation in the US and Europe*, pp. 1–582. Washington, DC: Resources for the Future.

12 Pesticides used in animal production

Key questions to consider

1 What laws and regulations protect people from the use of harmful pesticides?
2 Why do we allow minor amounts of pesticide residues to be in or on food?
3 How do we protect consumers of food from unsafe pesticide residues?
4 For food animals, what pests are being controlled through the use of pesticides?

Americans have embraced the use of pesticides. We dislike finding creepy, crawly little creatures in our houses or on our food. We employ copious amounts of pesticides to preclude infestations and control pests in our homes and yards. We also prefer fresh fruits and vegetables without blemishes, which often entails the use of pesticides.

Americans use about 1.1 billion pounds of pesticides a year, which equates to approximately 3.4 pounds per person (EPA, 2017). While most pesticides are used on agricultural crops, they are also used in animal production. An evaluation of meat products by the US Department of Agriculture (USDA) disclosed than more than one-third of the products contained pesticide residues (USDA, 2016b). It is difficult to avoid exposure to pesticides in the United States.

We recognize that pesticides are dangerous. To protect us and animals producing our food from the poisons in pesticides, we have turned to the government. Congress has enacted two major laws governing pesticide usage. The first is the Federal Fungicide, Insecticide, and Rodenticide Act (FIFRA). This law delineates provisions for the registration of pesticides and for the regulation of usage. The Environmental Protection Agency (EPA) oversees compliance with this law.

A second law, the Federal Food, Drug, and Cosmetic Act, is concerned with pesticide residues in or on food products. While we want to have our foods to be free from all pesticides, this is impractical. Pesticide-free foods are costly, and even organic foods may have low amounts of pesticide residues. Instead, we prevent situations in which pesticide residues present health risks to humans. The Food and Drug Administration (FDA) administers this law.

The production of animals for food involves the use of pesticides in raising animals as well as the production of grains used for animal feed. Producers treat animals with insecticides such as organochlorines, organophosphates, and car-bamates as well as with other chemicals to control pests and to avert diseases (Hildmann et al., 2015). We need to be concerned about whether uses are accomplishing their purposes of reducing disease, suffering, and discomfort for animals being treated. Simultaneously, we are concerned whether pesticide uses pose health risks to humans or denigrate ecological resources.

Producers of food animals use pesticides to control infestations of pests for financial reasons. Pests – including insects, lice, mites, ticks, and rodents – adversely affect animals' health and growth. While some infestations cause dis-comfort for animals, others may lead to sickness or death. Controlling pests is generally necessary for profitable animal production.

The evaluation of pesticides in animal production will commence with a dis-cussion of federal pesticide law, the registration of new pesticides, the cancelation of registered pesticides, and pesticide tolerances. With this foundation, we can look at the major pests for cattle, pigs, and chickens. Finally, we can note the issue of pesticides usage in producing feedstuffs and organic products.

Protection under FIFRA

To protect animals, people, and the environment from harmful pesticides, FIFRA requires every pesticide to be registered before it can be sold. Approximately 1,250 active ingredients are registered that are used in nearly 17,000 primary registered products. Tens of thousands of other distributor pesticide products are also sold. Compliance with the requirements of registration is costly, and many experimental formulations of pesticides never reach the market. For some of these formulations, unacceptable human health effects are the reason they never secure approval.

Registration under FIFRA requires information that supports a conclusion that a pesticide's projected usage is expected to be safe. Safety involves a pesticide performing its intended function without unreasonable adverse effects on the environment or harm to humans (Figure 12.1).

"Unreasonable adverse effect on the environment" is defined to require con-sideration of two categories of risks. First, the EPA considers unreasonable risks to humans or the environment based on the consideration of economic, social, and environmental costs and benefits of a use of the pesticide. The EPA performs a benefit–cost analysis that considers adverse effects and risks prior to the regis-tration of every pesticide. Benefits must be greater than associated costs for the pesticide to be registered under FIFRA.

Second, the EPA considers the human dietary risk from pesticide residues. Under the Federal Food, Drug, and Cosmetic Act, foods cannot contain pesti-cide residues that would expose persons to injurious amounts of the pesticide. If there is a pesticide residue in or on food, a tolerance is set so that the amount is not injurious. The EPA determines the maximum amounts of pesticide chemical residues that do not harm people and approves the use of pesticides so long as tolerances are met.

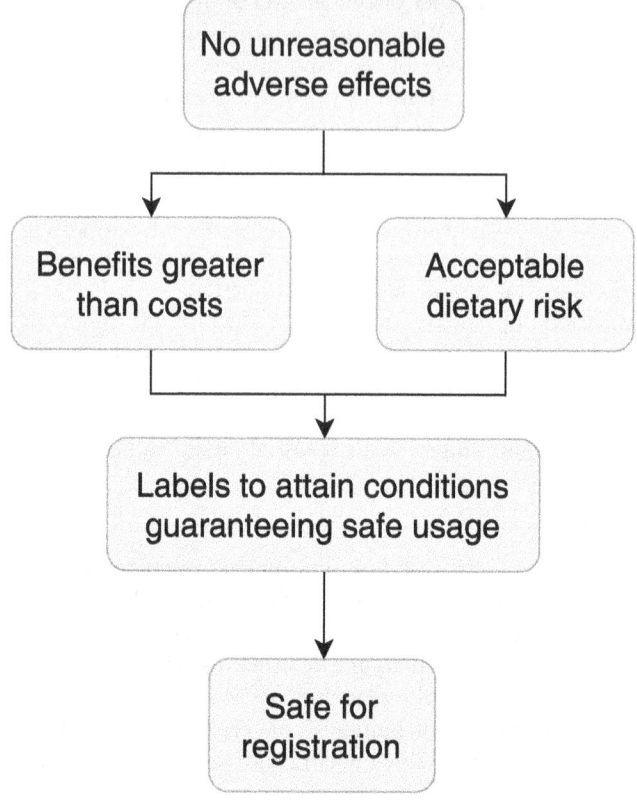

Figure 12.1 Regulatory framework for pesticide registration

Everyone using pesticide products knows that they have labels describing how to use them and warnings about dangers associated with usage. Registration requires pesticides to be labeled. In most cases, labels contain specialized instructions about how, when, and where the pesticide can be used. Labels have restrictions concerning concentrations, crops, and purchasers. Any person applying a pesticide in a manner that does not conform to its labeling requirements is violating the law.

Safety requirements for registration

While our laws require the EPA to keep us safe from dangerous pesticides, the agency is not conducting safety trials on pesticides. There are tens of thousands of registered pesticide uses and the EPA only has a limited number of employees. Rather, the agency relies on the registrant to provide this information. When a pesticide manufacturer seeks to register a new pesticide, it must submit the documentation showing that the pesticide's intended use is safe.

When a new pesticide is tested for safety, only the manufacturer and designated agents have access to the pesticide. All of the trials, studies, and evaluations of the safety of a new pesticide are done under the direction of the manufacturer. For a new pesticide that is similar to an existing pesticide, the manufacturer can submit safety studies that were used earlier for a similar pesticide.

This means that no governmental agency or a health advocate participates in providing evidence on the safety of a new pesticide prior to registration. If a registrant uses a faulty procedure, erroneous data, or misinterprets data concerning safety, the EPA will not be able to detect the falsehood prior to registration. Rather, the shortcomings of the testing procedure will not be detected until a later date, probably after people have been harmed.

Another concern is how firms conduct safety trials for a new pesticide. Obviously, we cannot rely on exposure studies involving humans. Instead, we are dependent on animal studies for determining the safety of a new pesticide. In controlled experiments, rats or some other animal are exposed to the new pesticide. They are observed for behavioral changes and, afterwards, are killed and their organs examined to determine whether exposure to the pesticide resulted in abnormalities. Most of the animal studies are for short durations, so we cannot discern the effects of a lifetime of human exposure. Additional safety margins are used to guarantee safety.

For pesticides used in conjunction with the production of food animals, concerns exist as to whether unhealthy residues are deposited in meat products. The USDA's Food Safety and Inspection Service tests meat products for 108 pesticides (USDA, 2016a). Obviously, the Service only samples a small number of products. If there is a violation, the violator is notified and the government evaluates the appropriate action to take. Repeat violators are posted on a weekly website. Of the 7,067 samples evaluated in a recent year, only five showed residues above safety limits (USDA, 2016b). However, 35 percent did display the presence of pesticide residues.

Cancelation of registration

As more information becomes available on the effects of pesticide usage, it sometimes becomes apparent that usage is accompanied by considerable harm. If the harm is severe enough, we want the pesticide's use to end. FIFRA delineates a mechanism to cancel a pesticide's registration if information discloses new risks or additional adverse effects that were not previously considered (Figure 12.2).

In addition, the FIFRA cancelation provisions allow any interested person to petition the EPA to cancel the registration of a pesticide. Under this provision, environmental and health groups have challenged registrations to terminate pesticide uses accompanied by too many damages. The burden is on the petitioner to show that the pesticide no longer qualifies for registration.

Under the FIFRA cancelation provision, the EPA considers the impacts on production and prices of agricultural commodities, retail food prices, and otherwise on the agricultural economy. This often involves options of adjusting

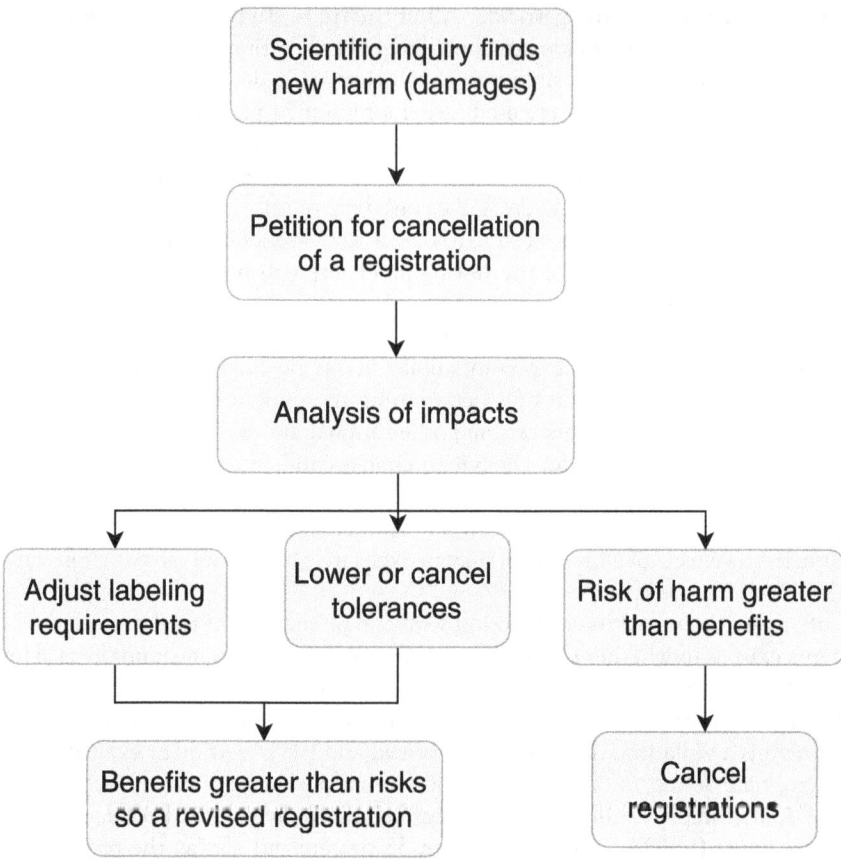

Figure 12.2 Regulatory framework for pesticide cancelation

labeling requirements to reduce usage. Tolerances may be lowered or canceled to reduce exposure. With these adjustments, the benefits from using the pesticide would be greater than the costs, so it could qualify for a revised registration.

The EPA may only cancel a registration when credible evidence shows the pesticide's costs outweigh its benefits. This most often involves new information available after the pesticide's initial registration showing additional health and environmental risks. The availability of an alternative pesticide may also support the cancelation of a more dangerous pesticide.

Pesticide tolerances

The use of pesticides in animal production may leave residues in or on products, water, soils, and the environment. Federal law sets maximum residue limits for pesticide chemical residues on foods called "tolerances." A tolerance is the maximum residue level of a specific pesticide chemical that is permitted in or on a specific human food. Tests for approximately 700 pesticide residues are conducted each year (FDA, 2018).

Tolerances are based on acceptable dietary risks to protect the health of persons ingesting food. Under the provisions of the Federal Food, Drug, and Cosmetic Act and FIFRA, the human dietary risk from pesticide residues on foods is considered and must be consistent with federal safety standards. A chemical pesticide residue can only be in or on a food product if it is below the tolerance required for a safe product.

The tolerance must be set at a level at which there is reasonable certainty that no harm will result from a person's exposure to the pesticide chemical residues in or on food (Figure 12.3). Each tolerance must consider the risk of pesticide chemical exposure of infants and children. Each tolerance must also consider the cumulative effects from exposure with other substances that have common mechanisms of toxicity.

Federal law recognizes three major categories of dietary exposure to pesticides. First, exposure may come from food. Second, exposure may occur from pesticides residuals in drinking water. Finally, exposure may come from the use of pesticides

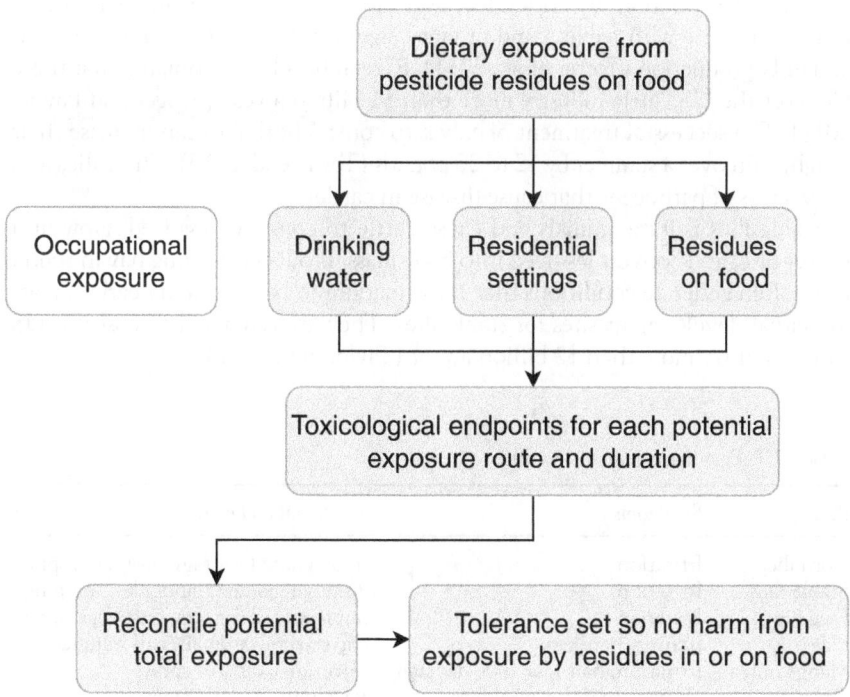

Figure 12.3 Establishing tolerances for pesticide residues in or on food

in residential settings. Dietary exposure does not include occupational exposure of persons using pesticides. The risk of dietary exposure is considered for a person's lifetime.

As might be expected, it is complicated to establish a tolerance for food products including meats. The EPA considers the toxicological endpoints for each potential exposure route and duration. The tolerance for food is established after consideration of the exposure pathways for drinking water and residential settings.

After a tolerance is set, a drinking water level of comparison is established so that people are not exposed to unsafe amounts of a pesticide. A drinking water level of comparison is a theoretical upper limit on a pesticide's concentration that is acceptable in drinking water in light of a person's total aggregate exposure to a pesticide. This level becomes important in situations where agricultural uses of pesticides are extensive.

Cattle

Livestock producers use pesticides to control insects, parasites, mites, and other harmful organisms. Common pests for cattle are listed in Table 12.1. Control of these pests is important for maintaining animal health and for helping the animals gain weight.

One pest is the horn fly, which feeds on cattle, horses, sheep, and dogs. Infestations interfere with feeding and produce significant reductions in weight gain and milk production (Torres et al., 2011). Researchers have estimated that these flies cost the US cattle industry more than $1 billion a year (Swiger and Payne, 2017). The successful treatment of calves to control horn flies can increase their weight gain over a summer by 12 to 20 pounds (Townsend, 2016). These flies also are vectors of pathogens that cause disease in cattle.

Stable flies irritate animals and cause cattle to consume less feed, grow at a slower rate, and convert less feed into body mass. Producers feeding hay in round bales often generate conditions that cause manure to be deposited nearby, creating larval development sites for stable flies. They are estimated to cost the US cattle industry more than $2 billion a year (Taylor et al., 2014).

Table 12.1 Pest management in cattle

Pest	Symptoms	Treatment and controls
Horn flies	Irritation	Impregnated ear tags, dust bags, sprays
Stable flies	Irritation	Remove manure, larvicides, trapping
Face flies	Irritation	Impregnated ear tags, dust bags, sprays
Lice	Itching, hair loss	Dip or spray animals, calf hutches
Mange mites	Irritation, hair loss, hide damage	Injections, dips, or sprays
Ticks	Damage to hides, weakness, anemia	Vaccination, dips, sprays

Face flies resemble house flies and gather around the eyes, mouth, and muzzle of cattle. They can transfer the pathogen *Moraxella bovis*, the causative agent of pinkeye. This is the most common eye disease for cattle and costs the industry an estimated $161 million per year (Maday, 2018).

Several species of chewing and sucking lice adversely affect cattle. Infestations may lead to reduce weight gain, milk production, and hide damage. While lice are not as damaging as flies, good husbandry practices can limit their numbers. Producers can prevent lice-infested animals from interacting with others and use individual calf hutches rather than pens to prevent infestation of other calves.

Various species of mites can infest cattle, causing sarcoptic mange. Mites contribute to poor weight gain, reduced feed conversion efficiency, decreased milk production, increased susceptibility to other diseases, and substantial devaluation of hides (Blutke et al., 2015). Sarcoptic mange is transferred by close physical contact between animals and is highly contagious.

Cattle tick infestations are a major problem in tropical and subtropical regions. Ticks reduce weight gain and milk production and transmit pathogens (Almazán et al., 2012). Vaccines are being used to control infestations, but more research is needed to develop additional protective antigens that would induce an immune response in cattle to species of ticks (de la Fuente et al., 2010).

One of the concerns is whether treatment of dairy animals with pesticides can result in unhealthy residues in milk. An analysis of 708 milk samples by the EPA detected just one pesticide (Flubendiamide) in one milk sample (USDA, 2018). The concentration of this pesticide was less than the established tolerance, so it did not pose a danger to human health. Sampling by the European Food Safety Authority found 92 percent of milk samples contained no pesticide residues, and no samples exceeded safety limits (EFSA, 2013).

Beef products in the United States generally do not contain harmful pesticide residues (USDA, 2016b). However, beef products tested in foreign countries often do contain pesticide residues. Meat tested in Jordan showed nearly 50 percent of the samples contained organochlorine pesticide residues (Ahmad et al., 2010). Testing of beef products in two African cities also showed significant organochlorine pesticide residues, but all within prescribed safety limits (Nuapia et al., 2016).

Pigs

Pigs are very susceptible to a number of diseases, including gastrointestinal parasites. Pests can transmit these diseases, so their control is important. Producers need to be on the lookout for signs of an infestation (Table 12.2).

Cockroaches may be an annoyance if allowed to infest a production facility. Flies are an annoyance but usually do not threaten the health of pigs; however, they may transmit disease agents, including *Escherichia coli* or *Streptococci suis*. Both cockroaches and flies may transmit organisms causing diarrhea.

Rats and mice are a problem at facilities raising pigs due to the abundance of feed. Rodents are often a reservoir for a variety of pathogens (Leirs et al., 2004).

Table 12.2 Pest management in pigs

Pest	Problem	Treatment and controls
Cockroaches	Spread pathogens	Insecticides, disinfect areas, separation of ages
Stable flies	Nuisance	Insecticides
House flies	Nuisance	Insecticides
Rats and mice	Transmission of mites and lice	Trapping, poison, rodent-proof construction
Mites and lice	Irritation, slower growth	Sprays and dusts

Mites and lice are a particular problem, as both rats and mice are carriers of these pests. Controls include rodent-proof feed containers, rodenticides, traps, and keeping dogs or cats.

Sarcoptic mange is a parasitic disease of the skin of livestock caused by a mite. The introduction of mites into a production facility may occur with new animals already infested or from wild mammals. Mange leads slower weight gains and greater piglet mortality.

Lice may be introduced into a herd by new animals or rodents. For most producers, management involves the use of a parasiticide. Lice can reduce feed conversion and hide quality, thereby reducing profitability.

To control pests and diseases, most pigs are produced at separate facilities for farrowing (birthing of piglets), nursery barns, and finishing barns. The large operations have stringent biosafety controls to limit people, rodents, and pests from gaining entry.

Pesticide residues in pork products produced in the United States are not a health issue. Tests by the European Food Safety Authority found that 97 percent of pork samples were free from detectable pesticide residues, and none of the detected residues exceeded tolerances (EFSA, 2013).

Poultry

Health issues involving poultry often concern the use of veterinary drugs – mostly antibiotics. However, the production of poultry is also accompanied by problems with flies, rodents, and mites. To control these pests, pesticides are often used.

Insecticides are generally applied as sprays, but sometimes are applied to birds. Care must be used not to contaminate water and feed supplies. The use of rodenticides must assure that birds are not affected by the poisons. This means that rodents must be controlled outside of the production area. Control measures might include minimizing points of access to buildings, elimination of nesting sites, and baiting and trapping programs. Failure to use rodenticides in a timely fashion can lead to major infestations and significant feed losses.

Northern fowl and red poultry mites can be serious parasites that need to be controlled. They cause economic losses by reducing egg numbers, egg weights, and hen feed conversion efficiency (Murillo et al., 2016). Mites also lead to increased mortalities. With the movement away from cages for egg-laying chickens, mite infestations have increased. Enriched cages, barn systems, and free-range housing systems offer environments favorable for mite infestations. Treatments usually involve careful use of chemicals or integrated pest management.

Residues of pesticides in poultry products are generally not a problem (Dervilly-Pinel et al., 2017). However, contaminated feed or housing containing hexabromocyclododecane has been found in eggs of laying hens (Dominguez-Romero et al., 2016). Moreover, meat and eggs from foreign countries may contain pesticide residues. A study in Jordan disclosed more than a quarter of their eggs and chicken meat may contain organochlorine pesticide residues (Ahmad et al., 2010).

Producing animal feedstuffs

Most livestock producers are dependent on feedstuffs that have been produced using pesticides. Insecticides and fungicides are used to control pests that might decimate crop yields. Herbicides – also known as weedkillers – are used to kill weeds that compete with food crops for sunlight, water, and nutrients. Insecticides are used to control insect pests in growing crops. The widespread use of these products is controversial due to potential human health risks and damages to the environment. Yet without the use of pesticides, grain prices would increase with corresponding increases of production costs for livestock.

Water contamination from herbicides used for producing corn and soybeans for animal feed is especially worrisome. Yet herbicide use is accompanied by positive environmental externalities, most importantly the replacement of cultivation. By eliminating or minimizing cultivation through no-till or conservation tillage, producers can markedly reduce erosion. Where soils are highly susceptible to erosion, no-tillage and strip tillage can reduce soil erosion by up to 90 percent (Zhou et al., 2009).

Another major use of pesticides is to protect stored grain. Because grain is often stored for six months or longer, care must be taken to prevent infestations of pests. Gaseous fumigant insecticides are used to control insect pests that may damage stored grain. Consumption of grain containing pesticide residues can lead to residues in animal products.

Organic animal products

Many people believe that organic animal products do not contain any pesticide residues. This is incorrect. Organic producers can and do use natural pesticides. Moreover, spray drift and pesticide vaporization from neighboring properties may result in small amounts of pesticide residues on organic crops.

Whenever a residue detected on the organic product did not originate from an intentional application of a prohibited substance, and the residue is present at levels that are less than federal regulations prescribe, the product can be sold as organic (USDA, 2012). This means that residues of synthetic pesticides may be on organic food items. Of course, the residues would be below any tolerance set by the EPA and so would not contribute to any health problems.

For animals, the main issue with organic production is whether the absence of use of a synthetic pesticide causes animals to suffer. Infestations of flies, ticks, lice, and mites can make animals uncomfortable. While natural products can reduce infestations, they do not always sufficiently control pests that adversely affect animal welfare. Producers may need to decide whether to allow an animal to suffer so it maintains its organic status or use a synthetic pesticide.

Foodwashing facts

1   We depend on our federal government to regulate pesticide usage so it does not harm people.
2   Pesticides are important in controlling disease in animals raised for food.
3   Pesticide residues may be on grains consumed by animals.
4   Animal products may contain pesticide residues but, generally, they do not a pose a health risk.

References

Ahmad, R., et al. 2010. Occurrence of organochlorine pesticide residues in eggs, chicken and meat in Jordan. *Chemosphere* 78, 667–671.

Almazán, C., et al. 2012. Control of tick infestations in cattle vaccinated with bacterial membranes containing surface-exposed tick protective antigens. *Vaccine* 30, 265–272.

Blutke, A., et al. 2015. Acaricide treatment prevents adrenocortical hyperplasia as a long-term stress reaction to psoroptic mange in cattle. *Veterinary Parasitology* 207, 125–133.

de la Fuente, J., et al. 2010. Identification of protective antigens by RNA interference for control of the lone star tick, *Amblyomma americanum*. *Vaccine* 28, 1786–1795.

Dervilly-Pinel, G., et al. 2017. Micropollutants and chemical residues in organic and conventional meat. *Food Chemistry* 232, 218–228.

Dominguez-Romero, E., et al. 2016. Tissue distribution and transfer to eggs of ingested α-hexabromocyclododecane (α-HBCDD) in laying hens (*Gallus domesticus*). *Journal of Agriculture and Food Chemistry* 64(10), 2112–2119.

EFSA (European Food Safety Authority). 2013. *The 2013 European Union Report on Pesticide Residues in Food*. Parma, Italy.

EPA (Environmental Protection Agency). 2017. *Pesticides Industry Sales and Usage: 2008–2012 Market Estimates*. www.epa.gov/pesticides/pesticides-industry-sales-and-usage-2008-2012-market-estimates.

FDA (Food and Drug Administration). 2018. *Pesticide Residue Monitoring Program Questions and Answers*. www.fda.gov/Food/FoodborneIllnessContaminants/Pesticides/ucm583711.htm.

Hildmann, F., et al. 2015. Pesticide residues in chicken eggs – A sample preparation methodology for analysis by gas and liquid chromatography/tandem mass spectrometry. *Journal of Chromatography A*, 1403, 1–20.

Maday, J. 2018. *Control Flies, Prevent Disease*. AG WEB, April 4. www.agweb.com/article/control-flies-prevent-disease/.

Murillo, A.C., et al. 2016. Northern fowl mite (Ornithonyssus sylviarum) effects on metabolism, body temperatures, skin condition, and egg production as a function of hen MHC haplotype. *Poultry Science* 95, 2536–2546.

Leirs, H., et al. 2004. Factors correlated with the presence of rodents on outdoor pig farms in Denmark and suggestions for management strategies. *Netherlands Journal of Agricultural Science* 52(2), 145–161.

Nuapia, Y., et al. 2016. Assessment of organochlorine pesticide residues in raw food samples from open markets in two African cities. *Chemosphere* 164, 480–487.

Swiger, S.L., Payne, R.D. 2017. Selected insecticide delivery devices for management of horn flies (*Haematobia irritans*) (Diptera: Muscidae) on beef cattle. *Journal of Medical Entomology* 54(1), 173–177.

Taylor, D.B., et al. 2014. Economic impact of stable flies (Diptera: Muscidae) on dairy and beef cattle production. *Journal of Medical Entomology* 49(1), 198–209.

Torres, L., et al. 2011. Functional genomics of the horn fly, *Haematobia irritans* (Linnaeus, 1758). BMC Genomics 21(105), 1–14.

Townsend, L. 2016. *Insect Control for Beef Cattle – 2016*. University of Kentucky Agriculture, Food and Environment. Lexington: University of Kentucky.

USDA (US Department of Agriculture). 2012. *2010–2011 Pilot Study; Pesticide Residue Testing of Organic Produce*. USDA National Organic Program, USDA Science and Technology Programs, November.

USDA. 2016a. *United States National Residue Program for Meat, Poultry, and Egg Products: 2016 Residue Sampling Plans*. Food Safety Inspection Service.

USDA. 2016b. *United States National Residue Program for Meat, Poultry, and Egg Products: FY 2016 Residue Sample Results*. Food Safety Inspection Service.

USDA. 2018. *Pesticide Data Program: Annual Summary, Calendar Year 2016*. Agricultural Marketing Service. www.ams.usda.gov/sites/default/files/media/2016PDPAnnualSummary.pdf.pdf.

Zhou, X., et al. 2009. Cost effectiveness of conservation practices in controlling water erosion in Iowa. *Soil & Tillage Research* 106, 71–78.

13 Selective breeding and animal cloning

Key questions to consider

1 Is the selection of breeding animals a recent development?
2 Why has selective breeding been important for the production of food animals?
3 Do we have a guarantee that we are not eating a food product from a cloned animal?
4 Do citizens in the United States and the European Union view animal cloning similarly?

Our ancestors have selected animals for food production since species of livestock were domesticated approximately 9,000 years ago. This involved breeding animals with desired characteristics and culling the remaining animals. The goal was to produce animals in a herd or flock that possess superior phenotypes that can be passed on to future generations. Producers often select more than one trait to advance the economic success of their production enterprise.

In the twentieth century, scientific advances made it possible to use records to evaluate genetic merit for pig, poultry, and cattle breeds. This facilitated alterations in the genetic makeup of animals through the selection of animals within a breed, between breeds, and through crossbreeding.

The most well-known selective breeding technology is artificial insemination. This involves collecting sperm cells from a male animal and artificially depositing them into females. To maintain the viability of the sperm cells, they are packaged in straws and stored in liquid nitrogen. It has been especially important in improving the genetics of dairy cows and the increased milk production of successive generations. About 80 percent of our nation's dairy cows are artificially inseminated, and about 94 percent of our sows (Colazo and Mapletoft, 2014).

Embryo transfer is a part of selective breeding. The use of multiple ovulation and an embryo transfer procedure enabled eggs from an animal to be collected, fertilized with sperm, and transferred to other females. In this manner, females with positive traits can have multiple progeny borne by others. With the advent of being able to cool and store embryos below the freezing point so they can be transferred to distant locations, embryos from the US are being used around the world.

More recently, cloning has been used for enhancing animal health and increased productivity. This technology is controversial but facilitates the increased production of food products to feed the world's growing population. Genetically modified crops are covered in the next chapter on genetic engineering.

Goals of breeding programs

Breeding programs have mainly focused on maximizing an animal's rate of weight gain. Selective breeding facilitates the production of the same quantities of meat products with fewer animals (Granola and Rosa, 2014). Due to selective breeding, pre-slaughter weights of livestock have been increasing since the 1920s. Cattle and hogs have more muscle and less fat. With animals gaining weight at a faster rate, they can be slaughtered at a younger age. Production practices have reduced prices of many meat products.

Some figures are illustrative. Since 1990, the average dressed weight for cattle has increased by 143 pounds, a 21 percent gain (USDA, 2016). For hogs, average dressed weights have increased 33 pounds in the same time period, an 18 percent gain. However, chickens had the highest increase in average live weight, a 40 percent gain. These weight gains have been accompanied by reductions in feed needed by food animals for the production of meat products.

Multiple ovulation and embryo transfer as well as *in vitro* embryo production allows the production of multiple progeny from a single female with desirable genetic characteristics (Paramio and Izquierdo, 2014). *In vivo* derived embryos and *in vitro* produced embryos, together with cryoconservation techniques, have increased animal productivity. Juvenile *in vitro* fertilization and embryo transfer is also contributing to genetic gains.

For dairy animals, the use of multiple ovulation and embryo transfers increased genetic gain in dairy breeding programs by up to 30 percent (Granleese et al., 2015). These technologies allowed shorter generation intervals in the breeding program. Recipient females give birth to offspring of the donor dam that have more desirable genetic traits.

For chickens, selective breeding has led to a fourfold reduction in the length of time to reach a broiler for meat production (Mignon-Grasteau et al., 2017). In the United States, broilers can be marketed at 47 days of age weighing over 5½ pounds each. This reduced growing period has been accompanied by a twofold reduction in the amount of feed consumed. Simultaneously, broilers have been bred to produce increased breast meat yield, as this is what Americans want for their chicken sandwiches and chicken nuggets. Selective breeding helps explain the decline in prices for chicken meat.

Goals other than weight gain are also important. One of the other goals has been to enhance disease resistance. Breeding has offered advantages over other disease control measures as a way to improve animal health and productivity. For farm-raised salmon, researchers have suggested breeding to help control sea lice (Gharbi et al., 2015). In Africa, researchers want to use breeding to enhance resistance to parasitism (Psifidi et al., 2017).

For cattle breeding, sex-sorted semen is available. After three decades of research, the collection of highly enriched populations of X- and/or Y-sperm are possible (Vishwanath and Moreno, 2018). For cattle producers, this facilitates breeding for more male cattle that have greater commercial value. However, in India the technology can be used to attain more female cattle because bulls are not needed for cultivation or transportation.

Cattle can also be bred to achieve more tender beef. IGENITY® gene profiles are commercially available to aid producers in the selection of cattle that better fit their needs (Magolski et al., 2013). Producers can decide what feature they want from a sire, including animal docility, calving ease, meat marbling, meat tenderness, and several other features.

Of course, breeding animals may be accompanied by negative consequences. A look at current chicken and turkey breeds suggests that selective breeding has led to species with major health problems. Today's flocks have musculoskeletal problems, male infertility, metabolic disease, and increased mortalities (Noll, 2013). Laying hens have behavioral, physiological, and immunological disorders. Yet, the industry is aware of these problems (Wang et al., 2018). In the past two decades, breeding programs have successfully addressed some of the issues that were adversely affecting the growth of birds and their yields.

Animal cloning

Another unconventional technique is also being used to foster the production of meat products: animal cloning (Centner and Petetin, 2018). This technology uses a somatic cell nucleus transfer technique to replicate animals. The main objective of animal cloning is to enhance animal genetic improvement. Cloning generates animals with desired traits for breeding so that their offspring can be used for producing food products. Other objectives include an increase of reproductive efficiency and resistance to disease (Centner and Petetin, 2018).

Animal cloning is not a new technique, as it has been used for nearly two decades. However, what has changed is that initially the harvested food products from clones did not enter the food supply. This may no longer be true. Today, there is no guarantee that the steak or hamburger you are eating did not come from a cloned animal.

While a number of animal species can be cloned, the most likely products you may consume from cloned animals are from bovine species: dairy products and beef. This may include milk, cheese, sausages, hot dogs, steaks, and hamburgers.

Cloning and food safety

Due to the concerns about safe food products, animal cloning was reviewed by the US Food and Drug Administration (FDA). In analyzing the issue of products from cloned animals, the FDA proceeded under the 1986 Coordinated Framework of Biotechnology. After appropriate scientific testing, the agency concluded that the products from animal clones and their offspring are similar to conventional animals.

The agency's conclusion meant that the safety of meat and milk of clones and their progeny compared with those from conventionally bred animals (FDA, 2008a). Thus, no new laws were needed, and products could be released to markets with identical requirements as conventional foods. However, the FDA recommended that products from cloned species other than cows, goats, and pigs should not enter the human food supply (FDA, 2008a, 2008b).

With the FDA's findings, the US Department of Agriculture (USDA) recommended that livestock producers should refrain from allowing food products from cloned animals to enter the nation's food supply. Under the USDA's direction, a voluntary moratorium in 2001 governed the sale of cloned meat products (USDA, 2008).

Yet, because this moratorium was voluntary, the government effectively handed producers complete control over cloned animals. When the offspring of cloned animals are sold to another farmer, there are no records of whether the animals were cloned. With the absence of such information, firms buying animals do not know whether they were clones or the offspring of clones. Meat packing plants without information on cloned animals cannot differentiate products from conventional, cloned, or descendants of clones. We must presume that products from cloned animals are being sold in supermarkets.

Cloning and animal health

Another concern accompanying animal cloning is the health of animals, both the health of the females providing the clones and their offspring. The FDA reviewed scientific evidence of the somatic cell nucleus transfer cloning technique to determine whether animal cloning and its derived foods were safe (US Center for Veterinary Medicine, 2008). Although the data showed a higher proportion of health risks for cloned animals, the FDA concluded that the risks did not qualitatively differ from natural breeding. While further testing was urged, the agency found "cloning poses no unique risks to animal health, compared to the risks found with other reproduction methods, including natural mating" (FDA, 2008a).

The issue of the ethical treatment of animals was also raised. However, the FDA does not have authority to address the ethics of animal cloning. Due to the omission of consideration of the ethical treatment of animals, the FDA's risk assessment has been subject to criticism by consumer and animal welfare groups.

Claims have been made that there are scientific studies showing genetic defects in clones being reproduced in clones' offspring. Moreover, consumer groups have argued that the FDA's risk assessment of animal health was compromised because it relied almost exclusively on data from biotech companies. In the absence of objective scientific studies, some consumers wanted the government to forbid animal cloning. Given public concerns, the USDA instituted its ineffective voluntary moratorium on the sale of cloned meat products in 2001 (USDA, 2008).

Cloning in the European Union

Regulators in the EU have not been as supportive of animal cloning (Petetin, 2012). The issue of animal cloning was initially regulated under the 1997 Novel Foods Regulation. Because cloned foods are considered as novel foods, preauthorization is required before they can be sold in the marketplace.

Two aspects of the EU's novel foods regulations may be identified. First, foods that are no longer equivalent to an existing food must be labeled. Labeling will inform consumers that something about the food is different. Second if a novel food is connected to an ethical concern, additional specific labeling may be required.

In 2018, a new Regulation created a centralized authorization system for novel foods and placed the European Food Safety Authority (EFSA) in charge of assessing risks connected with novel foods (EU, 2015). Under the Regulation, foods from cloned animals will need to be pre-approved and labeled for consumers.

However, under the EU's Regulation, foods derived from the descendants of cloned animals are not within the scope of the definition of a novel food. This means that products of descendants of cloned animals are not subject to the pre-market authorization, labeling, or traceability requirements of the Regulation. Since the descendants of clones will provide food products, consumers will not know whether their food came from a descendant of a cloned animal unless new requirements are imposed.

Future cloning oversight by the EU

Many Europeans are dissatisfied with their current regulations concerning food products from cloned animals. Consumer disgruntlement led the European Commission to submit two separate proposals dealing with animal cloning and derived foods in 2013. The first proposal would ban cloning practices in the EU (European Commission, 2013a). The second proposal would temporarily prohibit the placing on the market products from clones as well as their importation (European Commission, 2013b).

In 2015, the European Parliament supported the development of a new regulation to ban the cloning of all farm animals, their descendants, and products (European Parliament/Legislative Observatory, 2015). The Parliament felt that "animal cloning for food production purposes jeopardizes the defining characteristics of the European farming model, which is based on product quality, food safety, consumer health, strict animal welfare rules and the use of environmentally sound methods."

The European Parliament also noted its support for a regulation that would ban imports of animal clones, embryo clones, descendants of animal clones, semen, oocytes, and embryos collected or produced from animals for the purpose of reproduction, and food from animal clones and their descendants. The EU continues to examine more specific legislation covering the products of descendants of cloned animals.

Evaluation of animal health in Europe

In considering regulations regarding animal cloning, the European Commission asked the European Group on Ethics in Science and New Technologies to provide an opinion on the ethical implications of animal cloning with respect to food supplies (European Group on Ethics, 2008). Simultaneously, the EFSA was asked to produce an opinion on food safety, animal health, and environmental implications of live cloned animals.

The European Group on Ethics issued recommendations that were at variance with the American findings. The Group expressed doubts as to whether cloning animals for food supply is ethically justified. This was based on research that reported suffering and health problems of surrogate dams and animal clones. The European Group found no convincing arguments to justify the production of food from clones and their offspring.

Given the report by the European Group on Ethics, the issues of welfare and health of the surrogate dams carrying the cloned offspring were investigated by the EFSA (EFSA, 2012). This agency found a divergent situation than acknowledged by the US FDA. The EFSA found that the surrogate dams were affected by abnormal pregnancies. Moreover, the percentage of viable offspring born from transferred embryo clones was low at about 6–15 percent for bovine animals and about 6 percent for pigs (EFSA, 2012; European Commission, 2013b). Finally, the agency observed that calves and piglet clones had increased mortality within the postnatal and juvenile period. These data led the EFSA to conclude that the welfare of a significant proportion of clones was negatively impacted.

A comparison of the US and EU's evaluations

The American and European agencies evaluating cloning reached different opinions about animal cloning (Table 13.1). While both the FDA and EFSA found that food products from cloned animals are safe, they reached different conclusions on the health of cloned animals and surrogate dams. In the United States, it was found that the health of the dams of cloned offspring was not

Table 13.1 Comparison of the evaluation of food animal cloning in the US and EU

Ethical concern	United States	European Union
Health of dams carrying clones compared to other dams	Comparable	Inferior health
Welfare of cloned animals compared to non-clones	Comparable	Inferior welfare
Products of offspring of cloned animals	No differences	No differences but needs studying
Public opposition to cloning	Some	Significant

adversely affected, while the EU concluded they suffered by abnormal pregnancies. Regulators in the US felt that cloned offspring were comparable to other offspring, while the Europeans found evidence of low numbers of viable offspring.

Two other distinctions are important. The regulators in the EU felt more research was needed on food products from cloned animals and their offspring. Second, the public in the EU is supportive of regulating products from cloned animals, and some would preclude the sale of such products.

Consumer choice

While regulators have focused on scientific studies on the safety of products and the health of animals, perhaps governments ought to allow consumers to voice their opinions. Consumers may want information on products so they can choose whether they want to ingest products from cloned animals. While scientists can tell us answers about food safety and animal health, this does not address consumers' desire for information.

The scientific findings tend to ignore ethical and moral considerations that are important to a subset of consumers. These consumers want to consider animal welfare (Butler, 2009; Murphy, 2008). They are demanding that products from cloned animals be identified (Table 13.2). This is not that different from wanting to know whether a product is organic or whether a food product contains ingredients from a genetically engineered organism.

Studies on consumer attitudes have shown that a significant number of people want to avoid products from cloned animals. Whether a firm is selling products from cloned animals in the US, Europe, South America, or Asia, studies of consumers show preferences for products not from cloned animals (Aizaki et al., 2011; Brooks and Lusk, 2011; Schnettler et al., 2015). Yet, given the inability to trace the cloning history of some animals, any information on animal products from clones would probably only cover recently cloned animals.

Table 13.2 Summary of consumer preferences for products from non-cloned animals

Country	Findings	Researcher
United States	Approximately 43% of consumers are unwilling to consume meat and milk products from cloned animals	Brooks and Lusk, 2011
Japan	Approximately 40–50% of respondents were very uncomfortable about the consumption of cloned beef both before and after receiving additional information	Aizaki et al., 2011
Chile	Study participants preferred milk from conventionally bred animals	Schnettler et al., 2015

While regulators in the US seem to have obscured the issue of cloned animals, actions in Europe may again raise the public's consciousness. If the EU enacts a regulation banning the importation of products from cloned animals and their descendants, as has been proposed, a range of US dairy and beef products may be affected. The current mixture of conventional animals with cloned animals and their offspring raises questions about the ability of US beef suppliers to guarantee that their products meet the standard that they did not come from a cloned animal or an offspring of such an animal. Yet, it may also be impossible to confirm that a product came from a descendant of a cloned animal.

The enactment of the proposed EU regulation on cloning might create another trade controversy between the European Union and the United States. A cloning controversy would join the disagreement that currently exists over the existence of beta agonist residues in meat products (Centner et al., 2014). Given the EU has data suggesting that cloning is somewhat inimical to animal welfare, justification for labeling exists. Unfortunately, trade wars tend to be distasteful and a waste of resources.

Foodwashing facts

1   Due to selective breeding, our food animals are healthier.
2   The public remains uncomfortable with products from cloned animals.
3   The FDA has determined that food products from clones or their offspring are comparable to products from non-cloned animals.
4   In the US, there is no guarantee that we are not eating a food product from a cloned animal.

References

Aizaki, H., et al. 2011. Consumers' attitudes toward consumption of cloned beef. The impact of exposure to technological information about animal cloning. *Appetite* 57, 459–466.

Brooks, K.R., Lusk, J.L. 2011. U.S. consumers attitudes toward farm animal cloning. *Appetite* 57, 483–482.

Butler, J.E.F. 2009. Cloned animal products in the human food chain: FDA should protect American consumers. *Food and Drug Law Journal* 64, 473–501.

Centner, T.J., et al. 2014. Beta agonists in livestock feed: Status, health concerns, and international trade. *Journal of Animal Science* 92, 4137–4144.

Centner, T.J., Petetin, L. 2018. Divergent approaches regulating beta agonists and cloning of food animals: United States and European Union. *Society & Animals* 26(5), 1–20.

Colazo, M.G., Mapletoft, R.J. 2014. A review of current timed-AI (TAI) programs for beef and dairy cattle. *Canadian Veterinary Journal* 55, 772–780.

EFSA (European Food Safety Authority). 2012. Update on the state of play of animal health and welfare and environmental impact of animals derived from SCNT cloning and their offspring, and food safety of products obtained from those animals. *European Food Safety Authority Journal* 2794, 1–42.

EU (European Union). 2015. Regulation 2015/2283 of 25 November 2015 on novel foods. O.J. (L 327) 1–22.

European Commission. 2013a. COM(2013)892 final. *Proposal for a Directive of the European Parliament and of the Council on the Cloning of Animals of the Bovine, Porcine, Ovine, Caprine and Equine Species Kept and Reproduced for Farming Purposes*. http://eur-lex.europa.eu/legal-content/EN/TXT/PDF/?uri=CELEX:52013PC0892&from=EN.

European Commission. 2013b. COM(2013)893. *Proposal for a Council Directive on the Placing on the Market of Food from Animal Clones*. http://ec.europa.eu/food/food/biotechnology/novelfood/documents/cloning-2013-0433_app_en.pdf.

European Group on Ethics in Science and New Technologies to the European Commission. 2008. *Opinion No 23 on the Ethical Aspects of Animal Cloning for Food Supply*. Brussels, Belgium.

European Parliament/Legislative Observatory. 2015. *2013/0433(COD) – 08/09/2015 – Text Adopted by Parliament*, 1st reading/single reading. www.europarl.europa.eu/oeil/popups/summary.do?id=1401503&t=d&l=en.

FDA (Food and Drug Administration). 2008a. *Animal Cloning and Food Safety*. January. www.fda.gov/ForConsumers/ConsumerUpdates/ucm148768.htm, www.fda.gov/animalveterinary/safetyhealth/animalcloning/ucm055490.htm.

FDA. 2008b. *FDA Final Risk Assessment Finds Meat from Cloned Animals Safe*. www.beefusa.org/udocs/fdafinalriskassessmentfindsmeatfromclonedanimalssafe365.pdf.

Gharbi, K., et al. 2015. The control of sea lice in Atlantic salmon by selective breeding. *Journal of the Royal Society* 12(110), 0574.

Gianola, D., Rosa, G.J.M. 2014. One hundred years of statistical developments in animal breeding. *Annual Review of Animal Biosciences* 3, 19–56.

Granleese, T., et al. 2015. Increased genetic gains in sheep, beef and dairy breeding programs from using female reproductive technologies combined with optimal contribution selection and genomic breeding values. *Genetic Selection Evolution* 47(70), 1–13.

Magolski, J.D., et al. 2013. Relationship between commercially available DNA analysis and phenotypic observations on beef quality and tenderness. *Meat Science* 95, 480–485.

Mignon-Grastreau, S., et al. 2017. Genetic determinism of fearfulness, general activity and feeding behavior in chickens and its relationship with digestive efficiency. *Behavior Genetics* 47, 114–124.

Murphy, J.F. 2008. Mandatory labeling of food made from cloned animals: Grappling with moral objections to the production of safe products. *Food and Drug Law Journal* 63, 131–150.

Noll, S. 2013. Broiler chickens and a critique of the epistemic foundations of animal modification. *Agricultural and Environmental Ethics* 26, 273–280.

Paramio, M-T., Izquierdo, D. 2014. Current status of in vitro embryo production in sheep and goats. *Reproduction in Domestic Animals* 49(4), 37–48.

Petetin, L. 2012. The revival of modern agricultural biotechnology by the UK government: What role for animal cloning? *European Food & Feed Law Review* 6, 296–311.

Psifidi, A., et al. 2017. Genome wide association studies of immune, disease and production traits in indigenous chicken ecotypes. *Genetics Selection Evolution* 48(74), 1–16.

Schnettler, B., et al. 2015. Acceptance of a food of animal origin obtained through genetic modification and cloning in South America: A comparative study among university students and working adults. *Food Science and Technology, Campinas* 35(3), 570–577.

US Center for Veterinary Medicine. 2008. *Animal Cloning: A Risk Assessment*. Rockville, MD.

USDA (US Department of Agriculture). 2008. *FDA's Final Risk Assessment, Management Plan and Industry Guidance on Animal Clones and Their Progeny.* www.usda.gov/wps/portal/usda/usdamediafb?contentid=2008/01/0011.xml&printable=true.

USDA. 2016. *Overview of the United States Slaughter Industry.* National Agricultural Statistics Service.

Vishwanath, R., Moreno, J.F. 2018. Semen sexing – Current state of the art with emphasis on bovine species. *Animal* 12, S1, S85–S96.

Wang, Y., et al. 2018. Associations between variants of bone morphogenetic protein 7 gene and growth traits in chickens. *British Poultry Science* 59(3), 264–269.

14 Labeling genetically engineered plants and animals

Key questions to consider

1 What is genetic engineering?
2 Why do agricultural producers grow GE crops?
3 What do food safety experts say about the safety of foods from GE crops?
4 What new genetic engineering technologies for breeding animals are being used in the United States?

An issue that is important to many consumers of food products is whether the ingredients come from genetically engineered (GE) crops. These are crops that are produced from genetically modified organisms (GMOs). The US has embraced the production of GE crops, as approximately one-half of US crop-land is planted to GE corn (called maize in other countries), soybeans, or cotton (National Academies, 2016). In the rest of the world, approximately 3.7 percent of the agricultural land is used to produce GE crops (Canadian Biotechnology Action Network, 2015).

Consumer surveys indicate that an overwhelming majority of Americans would like foods to be labeled regarding the presence of ingredients from GE crops (called GE ingredients). Although no dangers have been identified as con-nected to food products with GE ingredients, consumers are interested in know-ing this information.

The US Department of Agriculture (USDA) has approved 195 GE crop events, but nearly all of these are currently not available for production. Cur-rently, only ten GE crops are being grown in the US. Since most of them were initially introduced in the 1990s (Table 14.1), Americans have been consuming GE foods for two decades.

Genetic engineering

GE crops were developed using a scientific procedure that introduces a new trait into the plant. Of course, humans have been selecting plants with new traits for millennia. Native Americans experimented with the development of superior corn, bean, squash, and melon plants for centuries, and subsequent European

Table 14.1 Introduction of GE crops in the United States

GE crop	Approved	Trait	Authorization
Alfalfa	1994	Insect resistant	Federal Register 59: 46030
Arctic apples	2015	Anti-browning	Federal Register 80: 8589–8590
Canola	1994;	Fatty acid production	Federal Register 59: 55250–55251
	1995	Herbicide tolerant	Federal Register 60(13): 4097–4099
Corn	1995	Insect resistant	Federal Register 60(119): 32299–32300
		Herbicide tolerant	Federal Register 60(158): 42443–42446
Cotton	1995	Insect resistant	Federal Register 60(179): 47871–47874
Papaya	1996	Virus resistant	Federal Register 61(180): 48663–48664
Potatoes	1995	Insect resistant	Federal Register 60(47): 13108–13109
			Federal Register 60(85): 21725–21728
Squash	1992	Virus resistant	Federal Register 57: 40632
Soybeans	1994	Herbicide tolerant	Federal Register 59(99): 26781–26782
Sugar beets	1998	Herbicide tolerant	Federal Register 63(88): 25194–25195

settlers continued with conventional plant breeding programs. These involved the selection of specimens with desirable traits and reproducing them by pollination. By selecting only desirable specimens, plant species with new genetic material were developed for widespread use.

Genetic engineering used for crops often involves isolating a gene in an organism and inserting it in another organism's genetic material. Thereby, genetic engineering is an extension of traditional plant breeding that employs technology and existing organisms to create a modified organism. Traditional plant breeding involves an indirect modification of genetic material, while genetic engineering involves the direct modification employing technology.

Some of the main uses of genetic engineering have been to create GE plants to resist insect pests, tolerate herbicides, resist viruses, and survive difficult growing conditions (Figure 14.1). Most of the GE crops currently being grown were developed to be herbicide tolerant. Some GE crops are both insect resistant and herbicide tolerant. The most extensively grown GE crops are corn, soybeans, cotton, and canola.

To control insect pests, genetic engineering was used to insert a pesticidal gene from a soil bacterium, *Bacillus thuringiensis* (Bt), into plant species that produces a protein that kills insects. The production of Bt corn and cotton crops has reduced the amounts of insecticides needed to control pests in these crops by an estimated 7.6 million pounds per year (Benbrook, 2012). Bt corn also lowers toxins in corn associated with mold from insect-damaged kernels.

For herbicide-tolerant GE crops, with the introduction of the *Agrobacterium* gene into a species, glyphosate herbicides can replace cultivation in controlling weeds. Reductions of labor and equipment in the production of corn, soybean, canola, sugar beet, and cotton crops have reduced costs. Conversely, the production of GE crops increased use of herbicides and the development of glyphosate-resistant weeds.

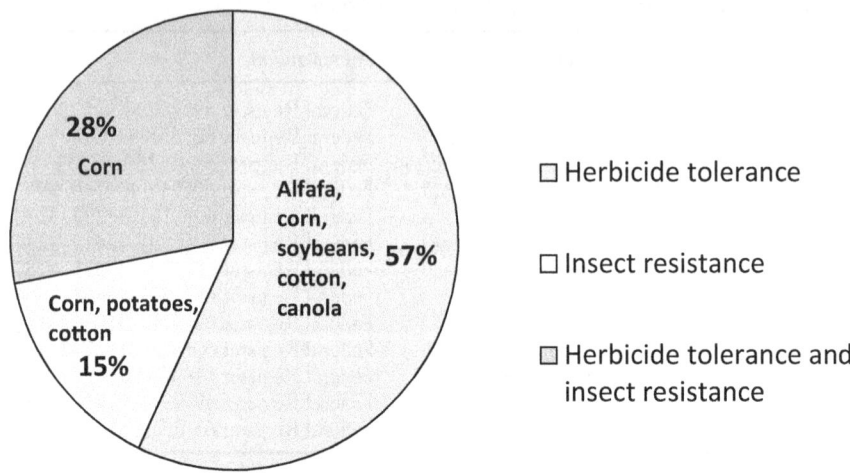

Figure 14.1 Differentiating percentages of the world's GE crops by trait*

* Canadian Biotechnology Action Network, 2015

Use in agricultural production and environmental concerns

Many of us are aware that US farmers are growing GE crops. In 2017, it was estimated that 92 percent of the corn produced in the country had at least one GE trait (USDA, 2017). For soybeans, 94 percent of acreage was a GE variety, while 95 percent of sugar beets produced in the United States were genetically engineered in a recent year. According to the International Service for the Acquisition of Agri-biotech Applications (2016), 78 percent of the world's soybean production and 64 percent of the world's corn production are GE crops.

The benefits associated with a GE crop vary substantially. For insect pests, GE crops should be considered when they make it economical for producers to cease using pesticides to control damages from insects. For weeds, producers with fields in which an herbicide is able to markedly reduce competition from weed species can benefit from growing GE crops. Researchers have estimated that the world has reduced fuel usage by 207 million gallons per year due to reductions in the need for spraying with insecticides and cultivating for weeds (Brookes and Barfoot, 2015).

GE crops have not been accepted in some areas of the world. Approximately half of Europe forbids the production of GE crops, as does most of Africa. Countries not planting GE crops include Russia, Ukraine, France, and Germany. However, producers in major corn, soybean, canola, and cotton producing countries are planting GE crops.

Several concerns exist about whether GE crops are bad for the environment. The most prominent is whether GE species have the ability to introduce their engineered genes into wild populations. GE corn, soybeans, and cotton crops have few sexually compatible weed species or naturalized plant species with which they could hybridize. However, GE alfalfa and canola have produced feral populations outside of cultivated areas, but no adverse environmental effects of gene flow from a GE crop to wild plant species have been documented (National Academies, 2016).

Researchers are investigating how to prevent crossbreeding and spreading of GE genes. One approach is to make the second generation of seeds sterile or dependent on a chemical for fertility. Another is to develop a GE plant that must be crossed with another GE plant in order for the offspring to contain the advantageous trait.

People are also concerned about the increased usage of herbicides. While this has accompanied the production of GE crops, there are also corresponding benefits. Some of the herbicides used on GE crops are less dangerous than herbicides used on non-GE crops. The use of herbicides reduces tillage and accompanying erosion, thereby reducing water pollution. Herbicide usage has lessened the costs of production of major food crops and so has helped reduce food costs.

Another concern is the susceptibility of non-target organisms to the gene product. What does the GE plant mean for beneficial insects and other living organisms? Testing is done prior to the release of a new GE plant to ascertain that there are no potential adverse consequences.

Allegations have been made that the development of GE crops is to blame for the loss of genetic diversity of many plant species cultivated for food. However, a study by the National Academies of Sciences, Engineering, and Medicine found that there was no relationship between GE crops and the loss of crop diversity for corn and soybeans (National Academies, 2016). Commercial agricultural production has led to the selection of only a few non-GE plant species because these enhance profitability.

Food products and safety

The large acreages of GE crops suggest a great number of food products may contain some GE ingredients. Nearly everyone in the US consumes GE ingredients unless they eat organic foods or avoid foods from supermarkets. GE corn and soybean ingredients account for a majority of GE foods consumed.

GE corn is used in high-fructose corn syrup and corn starch. Any food product containing either of these would contain GE ingredients unless costly efforts are made to source a product from non-GE corn. Small amounts of ingredients from corn are in many products, including breads, chips, cookies, cereals, fried foods, flours, pastas, and tofu dogs (Cummins and Lilliston, 2004).

GE sweet corn is available, although not widely grown.

Soybean oil accounts for 61 percent of our vegetable oil production. Nearly all of this soy-based oil comes from GE plants. GE ingredients would be in soy flour, lecithin, soy protein isolates, and concentrates. Products containing soy derivatives include tofu, cereals, veggie burgers, soy sausages, soy sauce, chips, ice cream, frozen yogurt, infant formula, sauces, protein powder, margarine, soy cheese, crackers, breads, cookies, chocolates, candy, fried foods, enriched flours, and pastas (Cummins and Lilliston, 2004).

Moreover, because most all of the canola grown is GE canola, anyone using canola oil is using a GE ingredient. Processed foods, chips, crackers, cereal, snack bars, frozen foods, canned soups, candy, bread, and hummus probably contain GE canola (Cummins and Lilliston, 2004).

Nearly 50 percent of the added sugar used in our country comes from GE sugar beets. This means that one-half of the added sugar we consume comes from GE plants, and products include cookies, cakes, ice cream, donuts, baking mixes, candy, juice, and yogurt. However, the production of sugar beets may change due to the refusal of some sugar buyers to purchase GE sugar. Many firms producing candy have decided to only use cane sugar to eliminate GE ingredients in their food products.

With these statistics, it is fair to say that most Americans have been consuming numerous products containing GE ingredients for years. Eighty percent of packaged foods in our supermarkets may contain GE ingredients (Beecher, 2014). It is estimated that a person consumes 193 pounds of GE food a year (Sharp, 2012). But persons eating organic foods can avoid GE ingredients, as organic rules forbid the certification of products from GE crops.

A common objection to GE ingredients is that they are not safe or may lead to health problems. Food items containing GE ingredients have been tested and retested hundreds of times to determine differences from non-GE crops. While a few studies claim problems, they have been discredited as lacking in appropriate scientific methodology. The experts cannot find any composition or health distinctions that are related to products from GM crops.

More specifically, studies conducted to determine whether GE crops might be related to increases in incidences of cancer or kidney disease concluded that the data do not support any association (National Academies, 2016). Another issue is whether food items from GE crops are related to current health issues of higher obesity, type II diabetes, or food allergies. No published research has shown any such relationship (National Academies, 2016). Furthermore, no published evidence supports the hypothesis that GE foods generate unique gene or protein fragments that might adversely affect humans (National Academies, 2016).

Even the European Union, with its reputed precaution for activities and food items that may cause harm, has found that GE crops are no more risky than conventional plant breeding technologies (European Commission, 2010). This conclusion was reached after analyzing research projects of more than 500 independent research groups.

Labeling GE products

The Food and Drug Administration (FDA) is in charge of regulating food safety. For more than two decades, the FDA declined to distinguish foods containing ingredients from GE crops. The approach was based on the policy that foods produced from GE crops did not require special labels because GE foods did not differ materially from other foods.

However, consumers have persisted in advancing the agenda that Americans have the right to know whether food products have GE ingredients. Due to consumer activism, state legislatures considered state labeling laws that would require the identification of products containing GE ingredients. A few states enacted GE labeling requirements, with Vermont's law being one of the most prominent. The law was scheduled to take effect in July 2016, but Congress decided that a common federal labeling standard was needed and enacted the National Bioengineered Food Disclosure Standard (GE Disclosure law) in 2016.

The GE Disclosure law requires labels for GE food products and preempts state and local GE labeling requirements. Rather than referring to GMOs or GE crops, Congress refers to "bioengineering," which includes a food

(A) that contains genetic material that has been modified through in vitro recombinant deoxyribonucleic acid (DNA) techniques; and
(B) for which the modification could not otherwise be obtained through conventional breeding or found in nature.

(US Code Service, 2018)

Under the GE Disclosure law, a food may bear a disclosure that it is bioengineered only in accordance with federal regulations. The Secretary of the USDA had two years to create and finalize rules implementing the law. The disclosure standard needs to determine the amounts of a bioengineered substance that may be present in food in order for the food to be called a bioengineered food. The Secretary also needs to establish a process for requesting and granting a determination regarding factors and conditions under which a food is considered a bioengineered food.

No specific food labels were established by the GE Disclosure law. Rather, the form of disclosure prescribed by the USDA's implementing regulations may "be a text, symbol, or electronic or digital link." A special provision governing food products derived from animals provides that a product cannot be considered a bioengineered food solely because the animal consumed feed produced from, containing, or consisting of a bioengineered substance.

Exploring the need for labeled products

A major argument for GE labeling has been that consumers have the right to know whether ingredients in their food came from GE sources. While this sounds reasonable, it may not be the question that we should be asking. There is a multitude of information that could be placed on food products, yet we cannot include every detail. From an economic and social perspective, the dividing line is whether the benefits of such label information are greater than the costs.

Proponents of labeling seem to feel that the benefits are worth the costs. However, they have not identified any meaningful benefits. While they claim health and environmental benefits, their claims are not substantiated by scientific research. All major research shows that GE products do not pose a health risk. Environmental risks are somewhere between nonexistent and highly speculative (National Academies, 2016).

As for the costs, no one knows what these will be. Different projections have been made based on assumptions that may not be realistic. But the costs will be more than the expense of affixing a label to each food product containing a GE ingredient. If manufacturers and retailers reformulate food products to respond to consumer objections to GE food, this will involve substituting non-GE ingredients (CAST, 2014). This may be costly.

The degree of negativity expressed by consumers in responding to products with GE ingredients is unknown. But companies that decide to reformulate food items with minor adjustments or substitutions to remove GE ingredients will incur costs in securing non-GE ingredients and, in some cases, product reformulation (CAST, 2014). Substitute ingredients will undoubtedly be more expensive and involve risks of inconsistent supplies.

New products will also involve decisions on whether to terminate existing products containing GE ingredients (McFadden, 2017). Some of these costs will be passed on to consumers, who have expressed some willingness to pay more for non-GE foods. The increased costs accompanying labeling will have a greater impact on low-income households than the general population.

A fundamental objection to imposing any costs for labeling products with GE ingredients is that it caters to the erroneous belief that these foods are bad (Sunstein, 2017). If GM labels produce a misleading signal that GE foods are unhealthy, the labeling requirement might be found to be counterproductive and offensive to free speech.

Legal issues connected to labeling

Although federal law requires some type of labeling of products containing GE ingredients, other provisions delineating rules concerning labeling may curb what can be required. Existing federal law precludes labels that are misleading. Under the Federal Food, Drug, and Cosmetic Act, misleading labels may constitute misbranding involving false or misleading statements. Labeling is not appropriate if consumers would be led to believe there is a difference between GE and non-GE products. In addition, the labels cannot infringe upon the First Amendment's freedom of speech.

Thus, courts may be asked to determine whether the attachment of a label noting that a product contains GE ingredients misleads consumers or whether consumers would interpret the label to mean that the product was not as safe or healthy as a non-GE product. Any labeling that would infer a health difference would be misleading and false, as there are no significant differences between GE and non-GE food items.

The labeling of products containing GE ingredients seems to be analogous to an earlier attempt to require dairy manufacturers to label milk products derived from milk produced by cows treated with rBST. Vermont enacted a mandatory rBST-labeling law in the 1990s and manufacturers objected. A federal circuit court found the law offended the US Constitution because consumer curiosity alone is an insufficient state interest to sustain a law that compels manufacturers to speak against their will (*International Dairy Foods Association vs. Boggs*, 1996).

Under this precedent, the government seems to lack a substantial interest for requiring GE products to be labeled, so mandatory labels may violate the First Amendment. Consumer curiosity is not a valid justification for forcing manufacturers to label products. In the absence of a justification for requiring labeling, manufacturers who do not wish to communicate information of GE ingredients should not be forced to do so. Voluntary labeling already exists and enables consumers to select products without GE ingredients.

Genetic engineering of animals

While traditional selective breeding programs have long been used to alter the genes of animals, the use of genetic engineering with animals was introduced in the 1970s (van Eenennaam et al., 2017). Genetic engineering commenced with pronuclear micro-injections of recombinant deoxyribonucleic acid (DNA). This involves exogenous DNA being injected into one-cell embryos that provide access to genetic variation not normally present in the target species. Thus, new transgenic animals are being developed (Laible et al., 2015). For large animals including cattle, pronuclear injections were not very efficient in developing offspring retaining the transgene (West and Gill, 2018).

A second genetic engineering technology is somatic cell nuclear transfer (SCNT). This technology allowed the development of transgenic animals, but most had low practical utility (West and Gill, 2018). The first genetically engineered livestock were introduced in 1985, and some noteworthy examples of genetically engineered animals are set forth in Table 14.2 (van Eenennaam, 2017). As may be noted, the inserted genes came from a number of species.

Scientists continued their research and developed additional technologies that would modify an animal's genome, the genetic material of the organism. Genome editing is a type of genetic engineering in which site-directed nucleases are used to precisely introduce a double-stranded break at a predetermined location in the genome (van Eenennaam, 2017).

Three assisted production techniques are being used for securing genetic changes (Ruan et al., 2017). Transcript activator-like effector nucleases (TALENs) are restriction enzymes that cut specific sequences of DNA. Clustered regularly interspaced short palindromic repeats (CRISPR) can be employed to confer resistance to foreign genetic elements and provide immunity. Zinc finger nucleases (ZFNs) can target unique sequences within genomes. These technologies can be used to truncate or knock out genes that are not wanted in an organism. They can also be used to allow insertion of desired changes to the genome.

Table 14.2 Genetically engineered food animals*

Species	Transgene	Origin	Trait/Goal
Cattle	Lysozyme	Human	Reduce mastitis (inflamed udders)
Cattle	SP110	Mice	Bovine tuberculosis resistance
Cattle	Prion Protein (PrP) shRNA	Made inoperative	Animal health
Chicken	alv6 envelope glycoprotein	Viral	Disease resistance
Chicken	LacZ	Fish	Animal health
Pig	Lysozyme	Human	Piglet survival
Pig	Mx, Iga, mouse monoclonal antibody (mAb)	Mice	Disease influenza resistance
Sheep	Visna resistance	Viral	Disease resistance

* van Eenennaam, 2017

These technologies may lead to hundreds of gene-edited food animals. Concerns exist about the regulatory burdens that will be placed on governmental agencies and the potential for off-target mutations in the genome (Ruan et al., 2017).

Yet, genetic engineering has tremendous practical applications for food animals. A genetically modified cow could be developed without horns that would make handling these animals safer and eliminate the need to dehorn the animals. Genome editing technologies can be used for the improvement of meat quality, such as breeding for more marbling and tender beef (Pimental and König, 2012). For pigs, genetic engineering might facilitate a species in which the male animals no longer have boar taint in their meat products.

Production of GE fish

In 2010, the FDA's Veterinary Medicine Advisory Committee met to consider a construct from AquaBounty Technologies, Inc. for a new genetically engineered AquAdvantage® salmon. The salmon contained genetic material from a Chinook salmon and an ocean pout causing the fish to produce quantities of a growth hormone to enhance weight growth.

In 2015, the FDA approved an application from AquaBounty Technologies to raise GE salmon for human consumption and sell their products in the United Sates. Prior to the FDA's approval, AquaBounty Technologies provided data and information in support of a New Animal Drug Application for a GE Atlantic salmon. Under the National Environmental Policy Act, an Environmental Assessment was conducted and the FDA issued a finding of no significant impact.

The FDA's approval included appropriate physical and biological containment measures to ensure the identity, quality, and purity of the animal lineage as well as to limit potential impacts on the environment. The all-female populations of triploid salmon are authorized to be produced at single facility on Prince Edward

Island in Canada. The embryos will be shipped to a single land-based grow-out facility in the highlands of Panama, where they will be reared to market size and harvested for processing into food.

The FDA's review ascertained that the inserted genes in the GE salmon species remained stable over several generations of fish, the salmon is safe to eat by humans and animals, the genetic engineering is safe for the fish, and the salmon meets the sponsor's claim about faster growth. Under US law, any deviation from the approved production conditions would mean the product was unapproved and adulterated under the Food Drug and Cosmetic Act.

Foodwashing facts

1 Scientific studies have not shown any health differences between food with and without GM ingredients.
2 GE labeling may mislead consumers into erroneously believing GE foods are inferior.
3 Voluntary labeling enables manufacturers to tell consumers that food products do not contain GE ingredients.
4 The public remains uncomfortable with products from genetically engineered plants and animals.

References

Beecher, C. 2014. Grocery manufacturers want foods with GMOs to be labeled as "Natural." *Food Safety News*, January 3.

Benbrook, C.M. 2012. Impacts of genetically engineered crops on pesticide use in the U.S. – The first sixteen years. *Environmental Sciences Europe* 24, 24.

Brookes, G., Barfoot, P. 2015. Environmental impacts of genetically modified (GM) crop use 1996–2013: Impacts on pesticide use and carbon emissions. *GM Crops & Food* 6, 103–133.

CAST (Council for Agricultural Science and Technology). 2014. *The Potential Impacts of Mandatory Labeling for Genetically Engineered Food in the United States*. No. 54, April.

Canadian Biotechnology Action Network. 2015. *Where in the World Are GM Crops and Foods?* Report 1, March 30. Ottawa, Canada.

Cummins, R., Lilliston, B. 2004. *Genetically Engineered Food: A Self-Defense Guide for Consumers*. New York: Marlowe & Company.

European Commission. 2010. *A Decade of EU-Funded GMO Research (2001–2010)*. Brussels: European Commission.

International Dairy Foods Association vs. Boggs. 1996. 92 F.3d 67 (US Second Circuit Court of Appeals).

International Service for the Acquisition of Agri-biotech Applications. 2016. *Pocket K No. 16: Biotech Highlights in 2016*. www.isaaa.org/resources/publications/pocketk/16/.

Laible, G., et al. 2015. Improving livestock for agriculture – technological progress from random transgenesis to precision genome editing heralds a new era. *Biotechnology Journal* 10, 109–120.

McFadden, B.R. 2017. The unknowns and possible implications of mandatory labeling. *Trends in Biotechnology* 35(1), 1–3.

National Academies. 2016. *Genetically Engineered Crops: Experiences and Prospects. National Academies of Sciences, Engineering, & Medicine.* Washington, DC: National Academies Press.

Pimental, E.C.G., König, S. 2012. Genomic selection for the improvement of meat quality in beef. *Journal of Animal Science* 90, 3418–3426.

Ruan, J., et al. 2017. Genome editing in livestock: Are we ready for a revolution in animal breeding industry? *Transgenic Research* 26, 715–726.

Sharp, R. 2012. *Americans Eat Their Weight in Genetically Engineered Food.* Environmental Working Group, Washington, DC. www.ewg.org/agmag/2012/10/americans-eat-their-weight-genetically-engineered-food#.WbXN8vN95pg.

Sunstein, C.R. 2017. Mandatory labeling, with special reference to genetically modified foods. *University of Pennsylvania Law Review* 165, 1043–1095.

US Code Service. 2018. Title 7, section 1639.

USDA (US Department of Agriculture). 2017. *Adoption of Genetically Engineered Crops in the U.S., Genetically Engineered Varieties of Corn, Upland Cotton, and Soybeans, by State and for the United States, 2000–17.* www.ers.usda.gov/data-products/adoption-of-genetically-engineered-crops-in-the-us.aspx.

van Eenennaam, A.L. 2017. Genetic modification of food animals. *Current Opinion in Biotechnology* 44, 27–34.

West, J., Gill, W.W. 2018. Genome editing in large animals. *Journal of Equine Veterinary Science* 41, 1–6.

Part IV
Marketing and social issues

15 Providing information on production practices and inputs

Key questions to consider

1   What animal production practices do consumers find objectionable?
2   What are governments doing to oversee labeling claims regarding animal production practices?
3   Why are marketers prohibited from using claims like "hormone free" or "antibiotic free?"
4   Why does our federal government offer verification programs regarding animal production programs?

The public's interest in healthy food extends to practices employed in animal production. Consumers want to know if animals were given hormones, treated with antibiotics, or raised outdoors. The food eaten by animals is also important to some consumers, as they want to avoid products from animals fed beta agonists, animal by-products, or grain. These consumers are willing to pay more for products from animals not receiving certain production inputs.

To inform consumers about the use or nonuse of production practices, vendors seek to append labeling information on meat and poultry products. By appending information about a production practice on a product label, the product can command a price premium. Part of this premium is paid to the producer to cover the additional costs that accompany the production practice. Whenever consumers are willing to pay more than the marginal cost of a specialized production practice, it becomes profitable for producers to adopt the practice to garner profits (McKendree et al., 2013).

Labeling products with information on production practices is challenging. The information should not only avoid false statements, but also should not be misleading. Without a clear definition about a production claim, there is a possibility that a label will be misunderstood by some consumers.

To protect consumers from false, fraudulent, and misleading product labels, governments have enacted labeling laws and regulations. The laws were adopted by various levels of governments to address problems and categories of concerns. In the United States, both Congress and state legislatures can enact legislation on some issues concerning livestock production and marketing. In the EU,

Commission regulations apply to all countries, but individual national governments can add to the restrictions. This means multiple laws govern the labeling of meat products.

Production practices

Subsets of consumers have decided they will try to avoid selected meat products due to health, welfare, ethical, and personal considerations. Many of the objections are based on the use of a production practice, and the 11 most frequently cited practices are listed in Table 15.1.

Some consumers are very deliberate in not purchasing products that are associated with one or more of these production practices. However, others are more flexible and try to avoid products that do not meet their standards. Together, the purchasing decisions of these consumers make it worthwhile for producers and vendors to develop certification and labeling programs that guarantee the authenticity of labeled products.

Hormones and antibiotics are perhaps the most objectionable inputs used in the production of animals (Centner, 2017). While the meat products from animals treated with hormones or antibiotics are considered safe and should not adversely affect humans, some consumers refuse to believe this determination. They seek products from animals that were raised without these inputs. The Congressional Research Service (2015) reported that about 30 animal growth-promoting products are used in the United States.

Four practices concern what the animals are fed. Two feed items that are objectionable to some consumers are beta agonists and animal by-products, including rendered items and fishmeal. These feed items are used because they enhance the profits of producers. However, despite governmental assurances that the animal

Table 15.1 Production practices eliciting consumer interest

Practice	Consumer preferences
Hormones	Animals produced without hormones
Antibiotics	Animals produced without antibiotics
Beta agonists	Animals not fed feed additives containing beta agonists
Feeding animal by-products	Animals not fed feed containing rendered animal by-products
Grass-fed	Animals were fed only grass, forage, or hay
Grain-fed	Animals fed grain prior to slaughter to enhance white marbling of meat
Produced outside	Free range, free roaming, or grass-fed animals
Crate- and cage-free	No cages for chickens, pigs, and veal calves
Humane practices	Certifications that administered procedures do not cause unnecessary pain
Organic	Certified compliance with the federal organic program
Vegan	Guarantee of no animal ingredients

products are safe to eat, consumers feel differently. They seek to avoid meat products from animals fed these items.

Consumers are also seeking more information on animals being raised on grass and forage without feed grains. While the organic label would include grass-fed products, some producers do not want to meet the organic requirements, yet want to raise and sell animals that were grass-fed.

Consumers choose grass-fed beef due to their disapproval of practices used in the production of beef at feedlots. Feedlots bring thousands of cattle together, creating problems with their welfare and manure. The use of grains for feeding these animals prior to slaughter means the production is supporting monocultures of genetically engineered corn with accompanying use of herbicides and erosion. Conversely, some consumers seek grain-fed beef from feedlots because they want a tender meat product with white marbling.

Three other subsets of consumers seek products from animals raised without crates and cages, or that were raised outdoors or naturally. These consumers object to chickens, pigs, and calves that are confined in unnatural conditions and suffer due to the lack of space.

A considerable number of consumers want to make purchasing decisions that support the humane treatment of food animals. This involves providing animals sufficient space, using anesthetics when performing procedures of castration or removing horns, and humane slaughter techniques.

National and international organic certification programs have delineated rules and regulations for the production of organic meat products. Producers raise animals in living conditions accommodating their natural behaviors, only feed 100 percent organic feed and forage, and cannot administer antibiotics or hormones. Certified meat products under this program are guaranteed to be organic. However, an exemption for small producers allows the sale of uncertified organic products in the United States (US Code of Federal Regulations, 2018, tit. 7).

Finally, a small number of consumers look for products that are guaranteed not to contain ingredients of animal origin. Vegans and persons with allergies and holding religious beliefs want to know whether products contain ingredients of animal origin that they wish to avoid.

US federal laws and regulations

Venders of meat products use labels denoting production practices because consumers will pay more for these products. The price premiums for labeled products provide a temptation for dishonest vendors to affix labels that are false or misleading. Congress responded to the problem of false and misleading labeling with legislation containing penalties and sanctions. Table 15.2 lists several major federal laws with labeling requirements and the agencies in charge of enforcement. Three of these federal laws allow states to adopt additional complementary provisions to oversee sales of food products.

For food marketing, the most important federal law is the Federal Food, Drug, and Cosmetic Act administered by the Food and Drug Administration.

Table 15.2 The federal regulatory framework for labels on meat products

Legislation	Enforcement	Complementary state law
Federal Food, Drug, and Cosmetic Act	Food and Drug Administration	Yes
Lanham Act	Competitors	Yes
Poultry Products Inspection Act	USDA	No
Federal Meat Inspection Act	USDA	No
Federal Trade Commission Act	Federal Trade Commission	Yes
Country of Origin Law	USDA (partially repealed)	No

This act was designed to protect the health and safety of the public by preventing the misbranding of foods. Misbranding includes statements on a label that are false or misleading, as well as situations where the information on a label is not prominently placed thereon or the label lacks the common or usual name of the food (US Code, 2012, tit. 21). Moreover, since the absence of information relevant to the issue may cause labeling to be misleading, misbranding also precludes labeling information that without further details might be expected to mislead.

A second federal law that oversees mislabeling of food products is the Lanham Act. Congress created a cause of action for unfair competition through misleading advertising or labeling that was separate and distinct from mislabeling under the Federal Food, Drug, and Cosmetic Act. The Lanham Act prohibits false descriptions of fact and misleading representations (US Code, 2012, tit. 15).

Two additional federal acts administered by the US Department of Agriculture (USDA) are important in overseeing permitted labels for meat products. The Poultry Products Inspection Act precludes misbranding while the Federal Meat Inspection Act prohibits false and misleading labels (US Code, 2012, tit. 21, §§ 457, 607). This means the federal government can initiate actions to prohibit the use of labels on meat products that mislead consumers.

The Federal Food, Drug, and Cosmetic Act and the poultry and meat acts do not allow consumers to sue for false and misleading labels. Because of these limitations, many state legislatures have adopted consumer legislation that enables consumers to sue marketing firms that use false or misleading labels.

However, both the Poultry Products Inspection Act and the Federal Meat Inspection Act contain provisions that preempt additional requirements by state governments (US Code, 2012, tit. 21, §§ 467e, 678). Under these provisions, state governments cannot enact any requirements concerning meat inspection and labeling. This means that the labeling of meat products is different from other food products. While a consumer may have a state law claim for a mislabeled food item regulated by the Federal Food, Drug, and Cosmetic Act's misbranding legislation, no state law mislabeling claims are possible for meat products.

The Federal Trade Commission Act grants the Federal Trade Commission authority to regulate labels and advertising. The commission can take action against "unfair or deceptive acts or practices in or affecting commerce" (US Code, 2012, tit. 15, § 45). Under this authority, the commission can proceed against unfair business practices, including false and misleading labeling and advertisements of food products.

USDA guidance

Under the authority provided by Congress in the Poultry Products Inspection Act and the Federal Meat Inspection Act, the secretary of agriculture has authorized the USDA's Food Safety and Inspection Service to develop and provide meat labeling guidance. Due to consumer and vendor interest in identifying production practices on meat products, the agency has published guidance, which is summarized in Table 15.3. The agency not only identifies production claims it approves but also identifies claims that are inappropriate because they cannot be proved.

Why can't products be labeled as "hormone free" or "antibiotic free?" Why can't producers say their cows were naturally raised? The answers can be found by conducting a careful analysis of federal law and the application of scientific facts. Hormones may be used for the production of cattle. However, because all beef products contain natural hormones, they cannot be said to be hormone free. To account for these hormones, the USDA limits labels to statements such as "raised without added hormones" or "no hormones administered."

An additional rule applies to products from pigs and poultry because federal regulations prohibit the administration of hormones to these animals. Any statement on a pork or poultry product saying "no hormones added" would be misleading, so needs to be accompanied by a statement saying that "federal regulations prohibit the use of hormones." Thus, information on hormones is rarely used on pork and poultry products.

Table 15.3 Marketing label information for animal products*

Topic	Approved production claims	Unprovable claims
Hormones	Raised without added hormones; no hormones administered	Hormone free, residue free, residue tested
Antibiotics	Raised without antibiotics; no antibiotics added	Antibiotic free, drug free, chemical free
Feed	Not fed animal by-products	
Feed	Corn fed or grain-fed	
Outdoors	Free range, free roaming, or grass-fed	
Raised naturally	None	Naturally raised, naturally grown
Organic	Certified organic	Organic or organically raised

* USDA, 2011, 2015a

For antibiotics, meat products cannot be labeled as "antibiotic free" due to inability to verify through an antibiotic-residue test that no antibiotics were administered to the animal (USDA, 2002). However, a private verification program may be used to establish documentation that a producer has not used antibiotics. The verification concerns the nonuse of a substance in the production of animals and allows products to be labeled as "no antibiotics added."

The USDA's Agricultural Marketing Service facilitated naturally raised labels on products through a Naturally Raised Marketing Claim Standard (USDA, 2009a). The standard was created to reduce confusion of multiple differing definitions for naturally raised livestock products being used by sellers. However, in January of 2016, the USDA withdrew these standards. The Agricultural Marketing Service realized that it lacked authority for this program.

The USDA's Agricultural Marketing Service had also facilitated a Never Ever 3 Program for verifying that the livestock providing products had never been administered hormones or antibiotics or fed animal by-products. This program was withdrawn in 2015. The nonuse of these inputs can be guaranteed under another USDA verification program (USDA, 2016d). Verification and certification programs also may be used for other production practices.

Organic products are certified by accredited certifying agents under the USDA's National Organic Program. The program uses the organic seal to denote certification of the production and marketing of products to guarantee they meet the qualifying requirements. Simply stating a product is "organic" or "organically raised" creates confusion because it only relates to the product itself. Under the USDA's National Organic Program, certification of the production process is also required to qualify for an organic label.

USDA verification programs

The USDA offers multiple programs for verifying production practices. The first is the USDA Quality System Assessment Program, which may be used for a variety of agricultural product marketing programs (USDA, 2017). Companies seeking verification under this program submit quality management systems that document, implement, and maintain assurance that products conform to the requirements of the applicable program. Each approved program is audited at least twice per fiscal year.

A second major verification program is the USDA Process Verified Program (USDA, 2015b). For this program, applicants use clearly defined, implemented, and transparent process points that are supported by a documented management system or an independent verification by a qualified Agriculture Marketing Service auditor. The International Organization for Standardization's ISO 9000 series standards serve as the format for evaluating documentation of management systems for the process verified program. This program facilitates labeling "no antibiotics ever" on meat products.

The Livestock Feeding Claims Audit Program is a third program offered by USDA to allow companies to establish feeding claims for their livestock (USDA,

2016b). This program can be used by companies to authenticate their livestock feeding claims for livestock and meat products. A company establishes the phases of production and marketing that are described through the scope of the program. Alternatively, a company may use the USDA Process Verified Program.

The USDA also offers a documented procedure for Quality Systems Verification Programs that are voluntary, user-fee funded, audit, and accreditation programs (USDA, 2016d). Under these programs, suppliers of agricultural products or services can secure independent third-party verification of defined processes and process points. The approval of a program is based upon the audit findings and the recommendation of the audit team. Fees are assessed pursuant to an approved hourly rate published in the US Code of Federal Regulations.

The USDA Grass Fed Program for Small and Very Small Producers was designed to create opportunities for small-scale livestock producers who would like to have their ruminant animals certified as grass-fed (USDA, 2016a). To qualify, the producer could only market fewer than 50 cattle each year. In 2016, the USDA withdrew the grass-fed claim for ruminant livestock and the meat products derived from such livestock.

USDA export verification programs

Meat exports are an important segment of the market. Because several important importing countries have different restrictions on what inputs can be used in producing food animals, the USDA's Agricultural Marketing Service offers several export verification programs. Under these programs, the US government gives assurances that animals providing meat products were raised without the offensive input. Table 15.4 lists four major programs that are used to provide assurances to foreign buyers that products meet foreign import requirements.

The USDA Export Verification Program for non-hormone treated cattle was developed to enable American producers to qualify for exporting livestock

Table 15.4 Significant USDA export verification programs for production practices

Issue	Program name	Method of verification for exports
Hormones	Non-Hormone Treated Cattle Program	USDA Quality System Assessment Program or USDA Process Verified Program
Antibiotics	Porcine Export Verification Programs	USDA Quality System Assessment Program or USDA Process Verified Program
Beta agonists	Never Fed Beta Agonists Program	USDA Quality System Assessment Program or USDA Process Verified Program
Animal protein	Animal Protein Free Verification Program for Poultry	USDA Quality System Assessment Program

products to foreign countries, especially the European Union, that do not want meat products from animals administered hormonal growth treatments (USDA, 2016c). Traceable animals, supplier evaluations, internal audits, and yearly external audits are some of the significant aspects of the program.

Beta agonists, animal feed ingredients that help animals develop more lean muscle, are opposed by some consumers. The use of beta agonists in the production of food animals is prohibited in many countries around the world. In response to the beef export market, the USDA adopted a verification program for animals never fed beta agonists (USDA, 2014). Companies use one of USDA's two major verification programs to guarantee to foreign purchasers that the meat products came from animals that were not fed beta agonists.

For antibiotics, the USDA's Process Verified Program sets forth a procedure under which producers have a documented management program in which antibiotics are not used (USDA, 2015c). Under this program, an official listing of approved USDA Process Verified Programs is available so that marketers and consumers can label and select products from animals raised without antibiotics.

The Food and Drug Administration has determined that protein derived from mammalian tissues for use in ruminant feed is a food additive (US Code of Federal Regulations, 2017, tit. 21, § 589.2000). In addition, to prevent the transmission of bovine spongiform encephalopathy, materials from cattle are prohibited in animal food or feed (US Code of Federal Regulations, 2017, tit. 21, § 589.2001). These regulations help safeguard our food supplies. However, foreign countries may need an additional guarantee for poultry products. Thus, the USDA administers the Animal Protein Free Verification Program for Poultry so that poultry products sold abroad are backed by a governmental assurance that unacceptable animal protein was not fed to birds providing the products being exported.

Private certification programs

For a number of production practices, private certifications are important. The certification of organic products by private certifying firms uses the USDA organic seal. Three private certifying organizations certify that animals providing meat products were treated humanely. The American Humane Association offers the American Humane Certified™ mark or label for products, the Animal Welfare Approved seal is available to guarantee that humane and environmental standards were followed, and a Certified Humane® certification for products is available.

The American Grassfed Association adopted grass-fed ruminant standards in 2009 and clarified these in 2016 (American Grassfed Association, 2016). Under these standards, animals are fed only grass and forage from weaning until harvest, are raised on pasture without confinement to feedlots, are never treated with antibiotics or growth hormones, and are born and raised on American family farms. Since the USDA withdrew its major program for grass-fed beef, this private certification is important in providing information of the production of cattle without grain.

Country of origin labeling

In 2002, Congress enacted legislation that would inform consumers as to the country of origin of certain agricultural commodities (Public Law 107–171, 2002). Writing the final regulations took considerable time, but in 2009 mandatory country of origin labeling (COOL) requirements were adopted by the USDA (USDA, 2009b). Various groups objected to the COOL regulations, with a major concern being whether the provisions would discriminate against imported meat products.

Animal producers and the suppliers of meat products in Canada were especially concerned as they felt that the COOL provisions resulted in less favorable treatment for Canadian beef, pork, and livestock. These concerns led Canada and Mexico to request the appointment of a World Trade Organization (WTO) dispute settlement panel to examine whether the COOL provisions violated international trade obligations set for by Articles 2.1 and 2.2 of the Agreement on Technical Barriers to Trade (WTO, 2012).

The analyses of the COOL provisions showed that compliance with COOL would add expenses in segregating products from imported livestock from the remainder of the production chain. Due to these expenses, the provisions created less favorable treatment for Canadian and Mexican livestock. The WTO Dispute Resolution Panel and Appellate Body found that COOL violated the Agreement on Technical Barriers to Trade.

The ruling seemed to leave open the possibility of providing some information on country of origin labeling, so the USDA attempted to amend its rules so that they would comply with international commitments. In 2013, the USDA released a labeling rule that required information on animals' origin at birth, production, and slaughter (USDA, 2013).

This 2013 rule was also challenged, and the WTO Dispute Resolution Appellate Body found the US regulations did not meet the country's WTO obligations (WTO, 2015). The Appellate Body recommended that the Dispute Settlement Body request the United States to bring its measures into conformity with its international obligations. In 2015, Congress repealed the offensive country of labeling provisions for beef and pork products (Public Law 114–113, 2015). COOL continues to apply to labeling of lamb, chicken, goat, and venison meat, certain fish, perishable commodities, and some nuts.

Foodwashing facts

1   Some consumers seek foods certifying the absence of a production practice.
2   Regulatory oversight is needed to prevent the false labeling of food products regarding animal production practices.
3   Numerous verification programs offer consumers assurances on how animals providing food products were raised.

4   Exports of meat products may be difficult, as many foreign countries have greater restrictions of the use of hormones, antibiotics, and beta agonists.

References

American Grassfed Association. 2016. *Grassfed & Grass Pastured Ruminant Standards.* Denver. www.americangrassfed.org/about-us/our-standards/.

Centner, T.J. 2017. Differentiating animal products based on production technologies and preventing fraud. *Drake Journal of Agricultural Law* 22(2), 267–291.

Congressional Research Service. 2015. *The U.S.-EU Beef Hormone Dispute.* Washington, DC. https://fas.org/sgp/crs/row/R40449.pdf.

McKendree, M.G.S., et al. 2013. Consumer preferences for verified pork-rearing practices in the production of ham products. *Journal of Agricultural and Resource Economics* 38, 397–417.

Public Law 107–171. 2002. Section 10816.

Public Law 114–113. 2015. Section 759.

US Code. 2012. Title 15, Sections 45, 1125; Title 21, Sections 457, 467e, 607, 678.

US Code of Federal Regulations. 2017. Title 7, Section 205.101; Title 21, Sections 589.2000, 589.2001.

USDA (United States Department of Agriculture). 2002. United States standards for livestock and meat marketing claims. *Federal Register* 67, 79552–79556.

USDA. 2009a. United States standards for livestock and meat marketing claims, naturally raised claim for livestock and the meat and meat products derived from such livestock. *Federal Register* 74, 3541–3545.

USDA. 2009b. Mandatory country of origin labeling of beef, pork, lamb, chicken, goat meat, wild and farm-raised fish and shellfish, perishable agricultural commodities, peanuts, pecans, ginseng, and macadamia nuts. *Federal Register* 74, 2658–2707.

USDA. 2011. *Meat and Poultry Labeling Terms.* Food Safety and Inspection Service.

USDA. 2013. Mandatory country of origin labeling of beef, pork, lamb, chicken, goat meat, wild and farm-raised fish and shellfish, perishable agricultural commodities, peanuts, pecans, ginseng, and macadamia nuts. *Federal Register* 78, 31367–31385.

USDA. 2014. *Quality Systems Verification Program (QSVP) Never Fed Beta Agonists Program.* Agricultural Marketing Service, QAD 1007 Procedure, March 11.

USDA. 2015a. *Animal Production Claims: Outline of Current Process.* Food Safety and Inspection Service.

USDA. 2015b. *USDA Process Verified Program.* Agricultural Marketing Service, QAD 1001 Procedure, October 26.

USDA. 2015c. *USDA Process Verified Program: Transparency from Farm to Market.* http://blogs.usda.gov/2015/12/07/usda-process-verified-program-transparency-from-farm-to-market/.

USDA. 2016a. *Grass Fed Program for Small and Very Small Producers.* Agricultural Marketing Service, QAD 1020 Procedure, January 12.

USDA. 2016b. *Livestock Feeding Claims Audit Program.* Agricultural Marketing Service, QAD 1040 Procedure, March 28.

USDA. 2016c. *USDA Export Verification (EV) Program Specified Product Requirements Non-Hormone Treated Cattle (NHTC) for the European Union.* Agricultural Marketing Service, QAD 1013 Procedure, March 28.

USDA. 2016d. *Quality Systems Verification Programs General Policies and Procedures*. *Agricultural Marketing Service*, QAD 1000 Procedure, August 25.

USDA. 2017. *USDA Quality System Assessment (QSA) Program*. *Agricultural Marketing Service*, QAD 1002 Procedure, June 12.

WTO (World Trade Organization) Appellate Body Report. 2012. *United States – Certain Country of Origin Labeling (COOL) Requirements*, p. 496, WT/DS384/AB/R, June 29.

WTO. 2015. *Recourse to Article 21.5 of the DSU by Canada and Mexico: United States – Certain Country of Origin Labelling (COOL) Requirements*. WT/DS384/AB/RW & WT/DS386/AB/RW, May 18.

16 Organic products

Key questions to consider

1 Who decides the criteria for organic products?
2 How can we be sure we're buying an organic product?
3 Do all products that are labeled as organic qualify for the organic seal?
4 Are pesticide residues acceptable in or on organic products?

Buying organic foods has become popular. With any visit to a major grocery store, we cannot avoid noticing signs for organic products. When we shop at a farmer's market, we see vendors touting organic products. During the last decade, sales of organic products have doubled. Some grocery stores only sell organic products. Organic products account for about five percent of total food sales. This is expected to increase. The organic market is clearly important for producers, marketers, and consumers.

Qualification for calling food products organic is based on legislation. In the United States, food products must meet the standards delineated by the National Organic Program (NOP), established by the Organic Foods Production Act of 1990. Organic crops and livestock must be raised in a production system that emphasizes protection of natural resources and plant and animal health.

The US Department of Agriculture's (USDA's) Agricultural Marketing Service is the designated federal agency overseeing the production and marketing of organic products. Beginning in 1990, the USDA developed a series of federal regulations defining the standards governing the sale of organic products. The provisions of these regulations were controversial, as different philosophies exist on what should qualify as "organic."

Complementing these standards is a series of guidance documents in the NOP Handbook that responds to numerous issues on the qualifications for organic production. In addition, the USDA has issued major guidance documents for organic crop and livestock producers and processors (Coleman 2012a, 2012b; Coffey and Baier, 2012).

The Organic Foods Production Act also established a public advisory board known as the National Organic Standards Board. This Board provides recommendations to the government on issues involving the production, handling, and processing of organic products under the NOP. For new issues, the Board may appoint ad hoc committees that make suggestions about the suitability of a proposed idea concerning organic standards.

Organic production

Organic agriculture is a production system based on the principle of sustainability, which considers conservation and ecological objectives. Organic production employs preventive management of pests, diseases, and predators without the use of synthetics or proscribed materials. For crop production, this involves the building of a healthy soil for plants. Producers build productive soils by adding organic matter, encouraging populations of soil microbes and soil invertebrates, and the selection of crops. To control pests, organic crop producers provide habitats for beneficial insects that control populations of harmful insects. Livestock producers provide good nutrition, sanitation, and a low stress environment to prevent illness and keep their animals healthy.

The US regulations for implementing the NOP delineate a complex compilation of standards that govern organic products. To understand what qualifies as organic, we can start with the National List of Allowed and Prohibited Substances. This list includes several prohibited inputs, materials, and processes; the most significant are listed in Table 16.1. Most consumers know organic food products do not come from genetically engineered species, and no hormones, antibiotics, or animal drugs are allowed. These prohibitions enunciate standards that delineate major production differences between organic and nonorganic food products.

Table 16.1 Standards incorporated in prohibitions of inputs, materials, or processes used for the production of organic food

Topic	Issue	Citation
Genetically engineered species	Not allowed	Baier, 2012
Hormones, antibiotics, and animal drugs	Not allowed for livestock production	§§ 205.237* & 205.238*
Pesticides	Synthetics not allowed	§ 205.105*
Sewage sludge	Not allowed on cropland	§§ 205.105* & 205.203*
Animal feed	No feed supplements of urea, manure, mammalian, or poultry slaughter by-products or ionophores	§ 205.237*
Irradiation	Ionizing radiation of organic materials or products not allowed	§ 205.105*
Strychnine	Prohibited in livestock production	§ 205.604*
Nonsynthetic substances	List of nine prohibited substances	§ 205.602*

* US Code of Federal Regulations, 2018

A broad range of other standards adopted under the NOP also regulates organic production. Three of these might be mentioned to highlight the scope of the controls. First, organic crop producers are not allowed to burn crop residuals. Rather, the organic matter should be incorporated into the field as a soil amendment.

Second, organic livestock producers cannot withhold medical treatment from a sick animal in an effort to preserve its organic status. Sick animals pose a problem because producers lose money if they administer medical treatment involving a prohibited substance that causes the animal to forfeit its organic status. The adopted policy is that animals should not suffer, and determination of this issue is made by individual producers. Third, producers cannot use lumber treated with arsenate that comes into contact with soil or livestock.

While the general prohibitions listed in Table 16.1 are helpful, the development of the organic standards for the NOP involves numerous controversies concerning various standards. A recent example involves whether crops produced hydroponically should qualify as organic crops. These crops are produced without soil. Some people contend that soil is an integral part of organic production, so hydroponic produce should not qualify for the organic appellation.

Criticisms exist on the current state of permissible synthetic substances that can be used under our organic rules. The erosion of strict organic standards for products such as eggs, milk, and grains is controversial, and groups such as the Organic Consumers Association continue to advocate stronger organic standards.

Exceptions to the rules

While the list of prohibitions attempts to provide definitive production rules to differentiate organic and nonorganic products, they have been found to be too strict. To facilitate the production of a wide variety of organic products, the regulations for the NOP contain a series of exceptions. The exceptions delineate provisions describing qualifications, substances allowed, and techniques to facilitate the production of organic products. Table 16.2 delineates some of the more significant exceptions.

Two general observations may be made about these exceptions. First, a considerable number of synthetic substances can be used in organic production. Examples include the use of synthetic methionine in the production of organic chickens to maintain healthy and productive poultry. Anyone who claims no synthetic substances are used in producing organic products is not familiar with the regulations.

Second, some of the exceptions overrule the prohibited inputs, materials, or processes set forth by the NOP's general organic standards. Because these exceptions are listed, they supersede the general standards. Organic products may contain pesticide residues if they are below a threshold and were not due to any action by the producer. Although synthetic pesticides cannot be used, growers can use seeds dusted with a fungicide.

Third, to allow dairy animals to qualify under the organic standards, regular dairy animals can be managed under organic principles for a year and afterwards

Table 16.2 Summarizing significant exceptions from the strict prohibitions for producing organic products

Topic	Issue	Regulation*
Synthetic substances allowed	List of 22 livestock additives and supplements with qualifications for livestock production	§ 205.603
Synthetic substances allowed	List of 37 synthetic substances allowed as ingredients in or on processed products	§ 205.605
Seeds	Use of nonorganic seed and planting stock when organic is not commercially available	§ 205.204
Seeds	Use of peracetic acid, hydrogen chloride, and chlorine materials	§ 205.601
Nonorganically produced products	Listing of 25 agricultural ingredients that can be used in processed organic products if an organic version is unavailable	§ 205.606
Pesticide residue	Unavoidable residual environmental contamination up to 5% of the pesticide's tolerance level	§ 205.671
Poultry	Under continuous organic management beginning no later than the second day of life	§ 205.236
Dairy animals	Under continuous organic management beginning no later than one year prior to the production of the milk	§ 205.236
Emergency pests	Prohibited substances permitted due to an emergency pest or disease treatment program	§ 205.672

* US Code of Federal Regulations, 2018

their milk qualifies as organic. The justification for these minor exceptions is to enable consumers to enjoy a wider variety of organic products at reasonable prices.

Certification through certifying agents

The NOP delineates a certification process for guaranteeing that products meet the standards set forth for organic products in the NOP. Certified products can be identified by the USDA organic seal (Figure 16.1). The NOP relies on approved private, foreign, and state certifying agents for overseeing the qualifications of products for the organic seal. If a product is not certified, it cannot make any organic claim on the principle display panel or use the USDA seal.

A producer or handler desiring to obtain certification adopts practices required for organic certification and implements an Organic System Plan (USDA, 2013). The plan delineates all inputs and ingredients to show that production and handling activities comply with the regulations. With this documentation, the producer or handler submits an application, plan, and fees to a certifying agent. The certifier will review the plan and schedule a visit to the site by an inspector.

Each site of an organic producer or handler is inspected to determine whether the applicant has the ability to comply with the regulations, verify that the plan

Figure 16.1 USDA's organic seal

accurately reflects the operation's activities, and ensure that prohibited substances are not being used or applied. The inspector may communicate any potential noncompliance observed and request additional information that may be missing from the plan. The report prepared by the inspector is reviewed by the certifier, and certification is only given if the operation is fully compliant.

Most certifying agents are directly accredited by the USDA. Eighty certifying agents are currently accredited and authorized to certify operations to the USDA organic standards. Of these, 48 are based in the United States and 32 are based in foreign countries. Twenty-one additional certifying agents are authorized through recognition agreements between the US and foreign governments. Each of these certifying agents is authorized to issue an organic certificate to operations that comply with USDA organic regulations.

Organic labeling

To announce to consumers that food products were organically produced, information is usually conveyed through a product label. We see these seals on most organic products that are labeled for sale in grocery stores.

The regulations adopted under the NOP set forth labeling requirements that apply to fresh products and processed products that contain organic ingredients. Organic nonfood products, such as feed for animals, must also be produced and processed in accordance with NOP standards. The regulations set forth certification provisions that provide oversight to ascertain that products meet the established criteria.

However, if you shop for organic produce at a local farmer's market, you probably will see organic products without the USDA organic seal. An exception exists for labeling products organic by small farmers selling less than $5,000 per year (US Code of Federal Regulations, 2018, tit. 7, § 205.101). This exception recognizes that certification costs may preclude small producers from seeking certification, so these producers are granted a waiver. They can sell organic products that meet the requirements set forth by the NOP but cannot use the organic seal. Under this exception, consumers can enjoy organic products from small producers but realize that the products are not certified as they are not using a USDA seal.

Organic labels may also be used on processed foods. The regulations for these foods become even more complicated in determining how the ingredients can be highlighted to garner attention.

Foods that are at least 95 percent organic may carry the organic seal as long as the nonorganic ingredients are not commercially available as organic. Food products that are at least 70 percent organic may be labeled as made with organic ingredients.

Nutrition

Many consumers believe that organic foods are nutritionally superior and safer. Considerable research suggests that for most organic crops, there are no systematic differences between organic and conventional crops (Jensen et al., 2013; Laursen et al., 2013). However, minor exceptions may exist. Leafy organic vegetables may contain more polyphenols (total) and phenolic acids in comparison to conventional ones (Kazimierckak et al., 2013). The same research found that plants from conventional production were more abundant in total flavonoids and carotenoids.

Moreover, it has not been proven that organic food improves human well-being or health (Jensen et al., 2013). A comparison of organic and conventional ready-to-eat breakfast cereals of similar types shows no significant values in nutrition under a recognized nutritional scoring system (Woodbury and George, 2014).

Another common assumption is that organic products are safer, with pesticide residues noted as the concern. Given US law and regulatory oversight, all food products sold in the United States should be safe. If there are residues of pesticides on a food item, it cannot be sold unless the residues are below a tolerance limit set by the Food and Drug Administration. The levels were established with large safety factors to guarantee that pesticide residues on food products do not cause health problems.

The major health issue regarding food items is whether organic production practices allow disease-causing microbes or pathogens such as *Escherichia Coli*, *Campylobacter*, *Listeria*, and *Salmonella* to become established. Food-borne illness data collected by the US government does not distinguish between organic and conventional foods, so no comparisons may be made from this data (Harvey et al., 2016).

However, in a study of chicken meat, researchers concluded that the likelihood of buying *Campylobacter* contaminated broiler meat was three times higher for organic broilers than for conventional broilers (Rosenquist et al., 2013). Another study of mixed crop and livestock farms showed greater frequencies of *Salmonella* contamination in poultry meat samples from organic farms (Peng et al., 2016). Other research suggests organic leafy greens may be more likely to be contaminated (Mishra et al., 2016). These studies suggest organic products are less safe than conventional food products.

Permitted pesticide residues

Consumers often assume that organic food products do not have any residues of pesticides. This is not true, as the regulations allow small quantities of pesticide residues on organic food from inadvertent or indirect contact from neighboring conventional farms or shared handling facilities (USDA, 2012). Given the difficulties in achieving a zero tolerance level for pesticide residues, regulations allow residues of prohibited pesticides up to 5 percent of federal tolerance levels (US Code of Federal Regulations, 2018, tit. 7, § 205.671). However, the residues cannot have originated from the operator's application of prohibited pesticides and the operator needs documented efforts to minimize exposure to residues coming from other sources.

Confusion about the ability of organic products to contain pesticides was noted in the *Johnson vs. Paynesville Farmers Union Cooperative Oil Company* (2012) lawsuit. In interpreting the federal organic regulations, the Minnesota Supreme Court observed that crops may not be sold as organic if they have a prohibited substance on them at levels greater than 5 percent of the Environmental Protection Agency's tolerance level for that substance. Furthermore, the requirement of having no prohibited substances applied to lands for three years needed to be interpreted in conjunction with the exception for prohibited substances. Under this interpretation, a field or crop contaminated by pesticide spray drift by some other person may qualify as organic if the residue levels are below 5 percent.

The importance of the exception for low amounts of pesticide residues is acknowledged by a USDA study. In a sampling of organic products, the USDA found that 43 percent had residues (USDA, 2012). However, the residue levels of 39 percent of these products were below the legal threshold. The study acknowledges the difficulties of producing many products with zero pesticide residues to underscore the need for the exception to facilitate organic production.

Some states have adopted provisions to help pesticide applicators know where crops sensitive to pesticide spraying are located. In these states, organic producers can list their organic fields on a sensitive crop locator to alert applicators and neighbors of organic production (Driftwatch, 2018). With this information, neighboring farms can employ extra caution when applying pesticides near an organic crop so that drift does not lead to the disqualification of the crop's organic status.

Organic milk production

Consumers want organic milk and milk products. Although not as common as fruit and vegetables, organic dairy products are prominent in many supermarkets in the United States. In some cases, insufficient milk supplies exist (Greene and McBride, 2015). The production and marketing of organic milk raise issues of whether exceptions are needed to facilitate increased production and whether consumers are being misled by false health claims.

Organic milk tends to be expensive to produce. Under the regulations of the NOP, the use of antibiotics and hormones are prohibited. Most synthetic chemicals used the production of crops providing food for organic animals are also prohibited. Furthermore, at least 30 percent of the dry matter food for organic cows needs to come from pasture during the yearly grazing season.

These limitations on inputs and feedstuffs increase costs and contribute to lower milk production by organic cows. One study found that organic cows averaged 13,600 pounds of milk a year versus 19,000 pounds for conventional cows (Greene and McBride, 2015). Organic milk often sells for almost double the price of conventional milk.

Several claims about health benefits of organic milk have been shared on social media. Some consumers feel that organic milk is more nutritious than conventional milk. Some have found that organic dairy products may contain higher protein, alpha-linolenic acid, total omega-3 fatty acids, and conjugated linoleic acid compared to dairy products derived from nonorganic systems (Hafla et al., 2013; Średnicka-Tober et al., 2016). Others have concluded that if there are differences, they are not significant (Erasmus and Webb, 2013).

Two other health claims concern the potential for antibiotics and estrogen in conventional milk that cause organic milk to be superior. However, under US law, no milk can be sold containing antibiotics that would harm people. Estrogen is contained in all milk and no significant differences in concentrations exist between organic and conventional milk (Schwendel et al., 2015). The health benefits of organic milk are overrated. Furthermore, since conventional milk is fortified with vitamin D, it offers a health advantage over organic milk.

Related policy issues

In considering the merits of consuming organic products, their anticipated relation to the need of inputs and amounts of land needed for food production might be considered. Organic products are often touted for being beneficial for reducing the need for inputs without noting that they may require additional land resources to produce equivalent quantities of food.

For energy consumption, organic production reduces reliance on inorganic fertilizers, pesticides, and other processed inputs that involve energy usage. Yet the result of forgoing these inputs is lower yields per acre for organic crops. While research varies, organic yields generally average about 80 percent of conventional yields (de Ponti et al., 2012).

Lower yields are due to a number of reasons, of which three are prominent. First, commercial fertilizers often provide additional nutrients that result in higher yields. Since sufficient nutrients may not be available to organic crops, yields are lower (de Ponti et al., 2012). However, for some organic crops, yields are comparable (Singerman et al., 2011).

Second, some organic produce and crops will be lost to pests. While pests threaten all crops, the nonuse of synthetic pesticides means there is a greater risk of an outbreak that reduces yields of organic crops. Conventional producers use pesticides because they lead to greater yields that translate into more economical production.

Third, production of some crops without herbicides will require other methods to control competition from weeds. If cultivation is used, it will involve additional equipment, fuel, and manpower. If weeds are not controlled, they lead to reduced yields (Larsen et al., 2014). Producers use herbicides to reduce costs per unit of food output.

The lower yields per acre for organic crops means more land needs to be cultivated to grow the same amount of food. This can involve machinery being driven more miles per unit of food, so organic production may use more energy. Due to the need for more land, an increase in organic food production may be accompanied by the use of fragile lands or clearing forests for agricultural production. Greater productivity on existing agricultural lands reduces the need to cultivate more acreage, thereby preserving lands for other uses.

Foodwashing facts

1   Governmental rules are needed to define organic food products.
2   Dozens of synthetic substances can be used under US organic production regulations.
3   Nutritionally, most organic products are similar to nonorganic products.
4   Organic food products can contain low levels of pesticide residues.

References

Baier, A.H. 2012. *Organic Certification*. Washington, DC: National Center for Appropriate Technology.

Coffey, L., Baier, A.H. 2012. *Guide for Organic Livestock Producers*. Washington, DC: National Center for Appropriate Technology.

Coleman, P. 2012a. *Guide for Organic Crop Producers*. Washington, DC: National Center for Appropriate Technology.

Coleman, P. 2012b. *Guide for Organic Processors*. Washington, DC: National Center for Appropriate Technology.

de Ponti, T., et al. 2012. The crop yield gap between organic and conventional agriculture. *Agricultural Systems* 108, 1–9.

Driftwatch. 2018. *Welcome to Driftwatch*. FieldWatch, Inc. https://driftwatch.org/about.

Erasmus, L.J., Webb, E.C. 2013. The effect of production system and management practices on the environmental impact, quality and safety of milk and dairy products. *South African Journal of Animal Science* 43(3), 424–434.

Greene, C., McBride, W. 2015. Consumer demand for organic milk continues to expand – Can the U.S. dairy sector catch up? *Choices* 30(1), 1–6.

Hafla, A.N., et al. 2013. Sustainability of US organic beef and dairy production systems: Soil, plant and cattle interactions. *Sustainability* 5, 3009–3034.

Harvey, R.R., et al. 2016. Foodborne disease outbreaks associated with organic foods in the United States. *Journal of Food Protection* 79(11), 1953–1958.

Jensen, M.M., et al. 2013. Comparison between conventional and organic agriculture in terms of nutritional quality of food – A critical review. *CAB Reviews* 8(45), 1–13.

Johnson vs. Paynesville Farmers Union Cooperative Oil Company. 2012. 817 N.W.2d 693 (Minnesota Supreme Court).

Kazimierckak, R., et al. 2013. The comparison of the bioactive compounds content in selected leafy vegetables coming from organic and conventional production. *Journal of Research and Applications in Agricultural Engineering* 61(3), 218–223.

Larsen, E., et al. 2014. Soil biological properties, soil losses and corn yield in long-term organic and conventional farming systems. *Soil & Tillage Research* 139, 37–45.

Laursen, K.H., et al. 2013. Multielemental fingerprinting as a tool for authentication of organic wheat, barley, faba bean, and potato. *Journal of Agricultural and Food Chemistry* 59(9), 4385–4396.

Mishra, A., et al. 2016. A system model for understanding the role of animal feces as a route of contamination of leafy greens before harvest. *Applied and Environmental Microbiology* 83(2), e02775–16, 1–20.

Peng, M., et al. 2016. Prevalence and antibiotic resistance pattern of *Salmonella* serovars in integrated crop-livestock farms and their products sold in local markets. *Environmental Microbiology* 18(5), 1654–1665.

Rosenquist, H., et al. 2013. *Campylobacter* contamination and the relative risk of illness from organic broiler meat in comparison with conventional broiler meat. *International Journal of Food Microbiology* 162, 226–230.

Schwendel, B.H., et al. 2015. Organic and conventionally produced milk – An evaluation of factors influencing milk composition. *Journal of Dairy Science* 98, 721–746.

Singerman, A., et al. 2011. *Price Analysis, Risk Assessment, and Insurance for Organic Crops. Center for Agricultural and Rural Development Policy Brief 11-PB6*. Ames: Iowa State University.

Średnicka-Tober, D., et al. 2016. Higher PUFA and n-3 PUFA, conjugated linoleic acid, á-tocopherol and iron, but lower iodine and selenium concentrations in organic milk: A systematic literature review and meta- and redundancy analyses. *British Journal of Nutrition* 115, 1043–1060.

US Code of Federal Regulations, 2018. Title 7, part 205.

USDA (US Department of Agriculture). 2012. *2010–2011 Pilot Study: Pesticide Residue Testing of Organic Produce*. www.ams.usda.gov/sites/default/files/media/Pesticide%20 Residue%20Testing_Org%20Produce_2010-11PilotStudy.pdf.

USDA. 2013. *Instruction: The Organic Certification Process*. NOP 2601. www.ams.usda. gov/sites/default/files/media/2601.pdf.

Woodbury, N.J., George, V.A. 2014. A comparison of the nutritional quality of organic and conventional ready-to-eat breakfast cereals based on NuVal scores. *Public Health Nutrition* 17(7), 1454–1458.

17 Locally grown products

Key questions to consider

1 Who decides what food products are locally grown?
2 Do locally grown products create environmental benefits?
3 Do specialized producers have advantages in producing food products?
4 Why do we support local food production?

A movement to eat locally grown food has been increasing in popularity. Many feel that our system of industrial agriculture is overly dependent on large quantities of fuel to transport food that puts an unnecessary strain on the environment. Not only is food moved great distances, but it is also heavily packaged at distribution centers, consuming additional resources. A major assumption by those who want to eat locally grown foods is that the transport of food damages the environment.

The term "locavore" is used to denote a person who eats local food. Many locavores will admit that they are worried about carbon emissions that are created by foods being transported long distances. Carbon emissions are objectionable because they contribute to climate change. A majority of local food is fruit and vegetables, so these are the focus of local food production. However, local products from animals also are often available at local farmers' markets.

Locavores also support local food production because they want to support their local communities. By buying food from neighbors, locavores augment the economic well-being of their communities. Arguments are also made that local food is fresher and more nutritious. But an analysis of local food production discloses that while there are benefits, in some cases, misinformation undermines the core belief that locally grown food reduces energy usage.

Defining local food

Unlike the term "organic," there is no legal definition nor consistent definition of what food can be considered "local." The term's use is highly variable, unregulated, and confusing. Some proponents define "local" using the mileage from where the food is produced to where it is sold. Others use existing boundaries, like state or county lines, to denote local production.

Most discussions of local food look at the number of miles that non-local food traveled. But, because local is undefined, consumers may not get what they expect when they buy a locally grown product. Some products are transported hundreds of miles and still called local. Products from a neighboring state may be precluded from being labeled local even though they were produced nearby. In a few cases, sellers may claim they are selling local products since the products were produced by a local farmer in a community hundreds of miles away.

Consumers assume that produce at farm stands and farmers' markets is grown nearby. This may not always be the case. In order to provide a variety of products that will attract consumers, some stands and markets have turned to produce suppliers. For example, some local markets sell bananas, as customers want to buy them along with local tomatoes and vegetables.

There is also confusion among consumers whether produce sold as locally grown is raised with sustainable methods or is organic. These are separate issues. Local does not guarantee sustainable production and does not mean organic. Food animals providing locally grown products may have received antibiotics and may have been injected with hormone supplements.

Food miles

Locavores look to "food miles" for guidance in supporting environmental quality. Food miles are the distances from where food is produced to where it is consumed. The view that food grown locally is good for the environment has become the selection criterion for these environmentally conscious food purchasers. Locavores assume that because local food has been transported a shorter distance, there is a reduction in carbon emissions compared to produce of the same type that has been transported longer distances.

This assumption is not always true. An example is growing tomatoes locally in a greenhouse. The energy usage of operating a greenhouse may be far greater than the energy consumed for transporting tomatoes from a more hospitable climate. It is more efficient to import food products that are expensive to grow locally.

Food miles do not consider all of the energy components of the input–output life cycle of a food item (Lewis and Mitchell, 2014). Energy usage accompanies the production of the food, transport to consumers, storage, consumer journeys to markets, and other activities related to getting food products to consumers (Table 17.1).

The evidence suggests that the energy consumed to get food to markets is severely overestimated in the argument for supporting local food production. Transportation of food has been estimated to make up only 12 percent of food's life-cycle greenhouse gas emissions, and only 5 percent of emissions are related to the delivery from the producer to the retailer (Weber and Matthews, 2008). It is estimated that 83 percent of the energy used in connection to the food we

Table 17.1 Energy consumption for producing and distributing food*

Transport	Energy usage	Percentage
Production of food	Land preparation, cultivation, pest control	83%
Farm to retailer	Truck, train, plane, boat	5%
Other transport to consumers	Truck or van	7%
Other energy needs	Temperature control, storage, consumer journeys to markets	5%

* Weber and Matthews, 2008

consume was used to grow the food. Because food miles only measure how far food has traveled, it is not an accurate gauge of energy consumption related to a unit of food available for consumption.

Energy inputs per unit of food

What consumers might want to consider is the total energy consumed in providing food products. We want to reduce energy usage per unit of food. This involves looking at all the energy used in producing and distributing food to the point of consumption. To assess energy usage, a fuller evaluation of the total life-cycle greenhouse gas emissions per unit of food output is needed.

If food can be produced in a more efficient manner in a distant region, the energy resources saved may outweigh the energy use to transport the food to market. This is why food production has evolved into large operations where favorable conditions allow food to be produced competently and competitively. These non-local specialized operations have efficiencies of scale. Since they have too many products to sell in local markets, they sell them regionally, nationally, and internationally. More energy may be consumed by the production of fruits and vegetables by many small operators than by specialized large operations in another region.

Another aspect of energy per unit of food involves the consumption of meat products. The production of livestock involves the use of more resources than the production of plant-based proteins (González et al., 2011). Because the production of meat products incorporates inputs that involve the use of energy, reductions of meat consumption can lower energy usage. Many vegetarians recognize that they are lowering their carbon footprint and contributing to reducing global warming by not eating meat products.

Production efficiencies

Although difficult to quantify, efficiencies can be identified for the production of fruits and vegetables (Table 17.2). Of course, efficiencies vary with different production units. Yet, these generalizations help explain why food from distant

Table 17.2 Comparison of projected production efficiencies for local and specialized
producers

Criteria	Advantage	More efficient
Distance to market	Transportation costs	Local production
Equipment used for production	Reduction in labor costs	Specialized producers
Equipment for processing and marketing	Reduction in labor costs	Specialized producers
Training for workers and knowledge	Reduction of risks	Specialized producers
Incorporation of new technologies	Reduction in labor costs	Specialized producers
Marketing information and management	Adjusting to markets	Specialized producers
Methods to reduce waste and losses	Secondary uses	Specialized producers
Harvesting technology and packaging	Reduction in labor costs	Specialized producers

locations is competitive. Specialized producers are able to adopt labor-saving technologies and spread their costs over large volumes of produce.

The distance to market is an efficiency that often favors local production. Food grown far from where it is consumed makes multiple journeys that consume energy. Yet, local production involves the transport of small quantities of food products from scattered locations, which consumes energy (Lewis and Mitchell, 2014). Specialized producers have an advantage in being able to haul large quantities of food in a single load.

All of the other efficiencies tend to favor specialized food production. Since food production uses the most energy, let's look at several features that may favor specialized production. The first involves equipment in preparing the land, planting, and controlling pests that reduce crop yields. Because of the large volume of fruits and vegetables grown at many specialized facilities, they can justify the use of equipment that reduces equipment and labor costs. This means they can produce food more cheaply.

Specialized production operations also tend to have more equipment for processing and marketing their products. While the initial investments for equipment are costly, over time the savings in labor costs means large producers have lower costs per unit of food. Centralized production, processing, and marketing systems can minimize the use of energy.

By specializing in the production of selected fruits and vegetables, firms managing large operations may acquire more knowledge about crops being grown and provide more training to workers. These features may help reduce risks of an event such as an outbreak of a disease that adversely affects a crop.

Large firms may also have a greater awareness of new technologies and can afford to incorporate technologies into their production practices. These technologies can help reduce spoilage, help preserve freshness, and reduce labor costs.

Specialized fruit and vegetable producers also garner marketing information that facilitates better management of their business operations. With this information, they may be in a position to reduce waste and losses. Initially, they are more likely to be able to market their fresh produce every day of the week, rather than only on days when a farmers' market is available. They also may have a market for produce that cannot be sold as fresh produce, such as a processing facility. Better information allows specialized producers to limit losses from an unusual weather event or an outbreak of a disease. They also may have markets for irregular or misshapen produce.

Specialized producers with large operations may be able to afford the latest harvesting and processing equipment. Economies of scale enable the costs of this equipment to be spread out over large volumes of produce. This equipment can reduce labor costs and facilitate the timely shipment to distribution centers or retailers. For example, equipment to cool produce immediately after it is harvested can help preserve its freshness.

While there are benefits related to local food production, they generally do not include the reduction of energy usage or production efficiencies. Specialized agricultural production involving large quantities of fruit and vegetables being grown in areas or regions has occurred because of economic efficiencies. By lowering production costs, fruits and vegetables from distant locations may cost less than locally grown products.

Taste and quality

Support for local food production is related to taste and food quality. These are good reasons to buy local fruit and vegetable products. Due to the quickness that producers can get locally grown food to the market, it is fresher.

Local produce is usually allowed to complete more of the ripening process on the plant instead of being removed while still unripe. Locally grown food is the closest that many consumers can get to eating ripe produce picked from their own backyard garden. Since nutritional value and taste both begin to decline shortly after ripe produce is picked, local produce has an advantage. It tastes better.

Local suppliers can grow unique regional cultivars. There are many fruits and vegetables that taste great but do not ship well. These foods are not available in the supermarket because they are too perishable. Other fruits and vegetables are difficult to package and ship. Local producers again have an advantage in producing these items. Many local growers do this well and have superb products.

However, consumers do not always eat locally grown food within hours of harvest. If they only go to the local farmers' market once a week, produce will not be fresh when it is consumed several days later. Consumers make tradeoffs between purchasing fresh produce often or going to the market (store) less frequently.

There are, however, two features that can detract from fresh local produce. First, what do local producers do with fresh fruit and vegetables that do not sell the day they are harvested or at the weekly farmers' market? These producers face a problem. If they continue to offer harvested produce for sale, it may no longer be fresh.

Second, what does a local producer do with produce that matures on days there is no farmers' market or on days there are not enough buyers who want the ripe produce? Zucchini need to be picked at least every other day or they will get too large and lose their flavor. Green beans picked once a week will be so large that they are tough and will not taste good. Producers selling at weekly farmers' markets may have a problem in deciding what to do with produce that is ready to eat on days it cannot be marketed.

Furthermore, shipped food has not always been in transit or at a grocery store for a long time. Produce from Mexico may be sold in New York 12–24 hours after it is harvested. If it is bought and eaten within a few days, it may be as fresh as produce purchased several days earlier at a local farmers' market.

Specialized producers may be able to use technologies and other processes designed to help preserve the nutrient content and taste of food. Packing practices at distribution centers include blanching, freezing, and the application of surface coatings. While we may object to some of these practices, they help maintain the quality of non-local produce.

Finally, locally grown processed food items do not have the advantages associated with fresh produce. Locally processed items are usually done in small quantities, so that they are more expensive. Second, low volumes of sales may mean items are not sold immediately. There also is inadequate justification for concluding locally processed products are of better quality. Because local processing may have fewer quality controls, locally grown products may be inferior.

Food safety

The safety of local foods is a controversial topic. Locavores often maintain that when consumers have face-to-face relationships with food producers, there is personal accountability for food's quality and safety. Yet, unsafe food products are usually the result of carelessness or not following recognized safety procedures (Godwin et al., 2012). While local producers are unlikely to be careless, they are more likely not to follow recognized safety precautions due to the inapplicability of governmental food safety regulations.

For example, a study of mixed crop and livestock farms in Maryland and the DC metropolitan area showed that chicken products at farmers' markets had a greater frequency of *Salmonella* contamination than at conventional retail markets (Peng et al., 2016). A similar finding had earlier been observed from a study in Pennsylvania (Scheinberg et al., 2013).

The real distinction in safety between local and non-local food is whether sellers of food products follow the safety requirements set forth in governmental food safety regulations. Does the seller have a food safety plan? Has the seller conducted a hazard analysis to identify and evaluate reasonably foreseeable hazards for foods manufactured, processed, packed, or held for sale to determine whether there are any hazards requiring a preventive control?

Under federal law, on-farm manufacturing and processing activities conducted by a small or very small business have been exempted from the rules on hazard analysis, risk-based preventive controls, and the supply-chain program (Code of Federal Regulations, tit. 21, part 117). Because of these exempted safety practices, locally grown food may be less safe.

The exemption from hazard analysis and risk-based preventive controls means that small producers do not need to have safety plans, hazard analyses, and preventive controls. If a seller of food products fails to oversee these food safety aspects, then local foods will not be as safe as products supplied by national and international firms. Because lapses of good practices are the cause of many food illnesses, the exemption from food safety requirements suggests that local foods may be less safe than other foods.

In a similar manner, if local producers process food items that include ingredients not raised on the farm, the resulting products may not be as safe due to the absence of a supply-chain safety program.

The federal exemption for small farmers was based on helping reduce their production costs and a presumption that state and local safety regulations would be in place to protect consumers. Yet, small farmers and food producers have also secured exemptions from safety provisions at the state and local government levels (Miller, 2015). Various food freedom and food sovereignty bills have been adopted by states in attempts to limit regulations over local foods. Most of these efforts diminish food safety regulations, suggesting that local food products might not be as safe as similar products offered by regulated sellers.

Our country's laws on food safety mean that national and international food companies cannot skimp on safety. They are obligated to follow detailed regulations that have been enacted to protect consumers. Moreover, large firms have considerable incentive to avoid a situation in which an unsafe food product would harm consumers. Any lapse by any major firm can have significant negative consequences on their sales, including their products that are free from safety issues.

Social benefits of local production

The choice to buy local food is not only related to energy usage and tasty food. Several social motivations are important in supporting local food production. Locavores want to support their local communities and enable consumers to engage in meaningful relationships with the producers of their food (Dobernig and Stagl, 2015).

The initial social benefit is that consumers may have some type of relationship with a producer. Some people want to know who has produced their food. Their trust of these producers leads them to buy locally. This often is related to the belief that local producers will take care of both the environment and their community. Local producers are deemed to practice good husbandry practices and take care of their land. This generally is true, as local producers enjoy what they are doing and want their efforts to be acknowledged by their neighbors.

Keeping local farms under cultivation is a second reason to support buying locally. By encouraging local agriculture, consumers help preserve the use of farmland for agricultural and recreational use and wildlife habitats. Eating locally grown food aids neighbors who are hard-working farmers, stewards of the land, and preservers of the local farming community.

Social interaction with local farms is also an important educational opportunity for families with young children. Outings to pick your own operations and agritourism venues are great activities for families. Regional economic health and social connections to local producers are important reasons for preferring locally grown food.

Favoritism for local foods is not beneficial for some of the poorer countries of the world. Their farmers may work hard to produce low-cost produce that can be sold in distant markets. If the marketing opportunities for these products are reduced, it may make it difficult for farmers in developing countries to improve their conditions. Legislation promoting local food in ways that discriminate against products from other markets is contrary to global trade law.

Summarizing the merits of local production

The current movement toward local food has been built on assumptions concerning energy savings and support for local businesses. While some of the assumptions are reasonable, others are not supported by evidence or practices. Table 17.3 summarizes seven criteria that have been presented on the advantages of local food production. Because most energy used in the production and delivery of food to markets is used for producing food, and economies of scale favor large producers, the energy usage per unit of food varies. No clear preference may be identified.

However, in the efficiency of food production, large producers have advantages. Labor-saving equipment and technologies facilitate the low-cost production of large quantities of produce. Economies of scale mean that low-cost food items can be grown at distant locations.

Local food production may reduce spoilage of food and provide more flavorful and better quality food products for consumers. Eating local foods during their production seasons also makes sense. These facts support local production. Yet, if local producers are only able to market products once a week, they also may have a problem with spoilage.

Table 17.3 Comparison of advantages (positive) of local and specialized producers

Criteria	Local	Specialized
Energy usage per unit of food	Varies	Varies
Efficiency per unit of food		Positive
Product spoilage	Positive	
Taste and quality	Positive	
Safety: comprehensive program in place		Positive
Variety of food products and culinary experiences		Positive
Social: community fellowship	Positive	

Food safety favors non-local food products. This is because local producers have been exempted from significant food safety regulations and may not engage in procedures that would prevent a situation that causes an illness related to food. The comprehensive safety programs that are obligatory for specialized producers suggest their products will be less likely to cause a health issue.

Importing food items to an area allows consumers to enjoy a wider variety of products. This facilitates an enhanced variety of culinary experiences. Finally, local food production is good for communities. People meet their neighbors and learn how food is produced. Local production can keep farmland in more extensive production with positive spinoffs for the community.

Agritourism

Related to local food production is agritourism. Farmers and others are entering the tourism industry to supplement existing income sources. Agritourism is an increasingly important portion of local agricultural activities that encourages and helps local enterprises maintain economically viable businesses. For persons owning a farm or ranch, the increased revenue stream from agritourism can augment their income to keep their operations solvent. For others, it's a new business enterprise that may no longer be a farm.

Agritourism may be defined as farming-related activities carried out on a working farm or other agricultural settings for entertainment or education purposes (Chase et al., 2018). Yet some go further and maintain agritourism includes non-working farms, farmers' markets, outdoor recreation, and agricultural fairs. Under the 2012 US Census of Agriculture, agritourism sales and recreational services contribute more than $2 billion per year to farm income (Chase et al., 2018).

In some cases, attracting school kids and tourists is the focus of an operation. At the largest agritourism site in the country, kids are able to watch pigs being born and manure move through a system of anaerobic digestors (Vrabel, 2017). Agritourism activities include farm tours, pumpkin harvesting and painting, orchard tours, learning about farm machinery, Halloween parties, hay rides, wine tours, petting zoos, hunting for a fee, fishing for a fee, horseback riding, farm vacations, pick-your-own operations, camping, craft shops, country stores, roadside stands, farm museums, nature trails, picnic areas, and children's day camps.

Given the benefits of agritourism on an area's economy, governments are developing broader state legislation for agribusiness. In some cases, they are providing training sessions to help farmers learn how to structure an agritourism operation and make business decisions. Agritourism compliments local food production in supporting local communities.

Foodwashing facts

1   Local foods tend to be fresher and have a better taste.
2   Distances food products travel to market are not directly related to total energy usage.

3   Local foods may be less safe due to exemptions from food safety regulations.
4   Supporting local food production is good for communities.

References

Chase, L.C., et al. 2018. Agritourism: Toward a conceptual framework for industry analysis. *Journal of Agriculture, Food Systems, and Community Development* 8(1), 13–19.

Code of Federal Regulations. 2017. Title 21, part 117.

Dobernig, K., Stagl, S. 2015. Growing a lifestyle movement? Exploring identity-work and lifestyle politics in urban food cultivation. *International Journal of Consumer Studies* 39, 452–458.

Godwin, S.L., et al. 2012. Consumer response to food contamination and recalls: Findings from a national survey. *Journal of Food Distribution Research* 43(1), 17–23.

González, A.D., et al. 2011. Protein efficiency per unit energy and per unit greenhouse gas emissions: Potential contribution of diet choices to climate change mitigation. *Food Policy* 36, 562–570.

Lewis, M., Mitchell, A.D. 2014. Food miles: Environmental protection or veiled protectionism. *Michigan Journal of International Law* 35, 579–636.

Miller, S.R. 2015. A coordinated approach to food safety and land use law at the urban fringe. *American Journal of Law & Medicine* 41, 422–446.

Peng, M. 2016. Prevalence and antibiotic resistance pattern of *Salmonella* serovars in integrated crop-livestock farms and their products sold in local markets. *Environmental Microbiology* 18(5), 1654–1665.

Scheinberg, J., et al. 2013. A microbiological comparison of poultry products obtained from farmers' markets and supermarkets in Pennsylvania. *Journal of Food Safety* 3(3), 259–264.

Vrabel, J. 2017. New crop. *Indianapolis Monthly*, pp. 53–57, August 2017. www.indiana polismonthly.com/features/new-crop-fair-oaks-farms/.

Weber, C.L., Matthews, H.S. 2008. Food-miles and the relative climate impacts of food choices in the United States. *Environmental Science & Technology* 42(10), 3508–3513.

18 Animal waste management

Key questions to consider

1 Why are Americans becoming ill from foodborne illnesses?
2 Why are we concerned about animal waste?
3 Why is only animal waste from some farms regulated under the Clean Water Act?
4 Do governments do a good job at enforcing animal waste regulations?
5 Can common law legal actions be brought against producers who cause health and environmental problems?

With the concentration of animals in the United States at large production units and in regions with a competitive advantage, the disposal of their waste can cause problems. A majority of the beef, dairy cows, hogs, chickens, and turkeys are raised at 14,000 to 20,000 large farms. Their manure, urine, and process wastewater can overwhelm the capacity of nearby fields to assimilate their nutrients as a fertilizer. Some animal waste ends up in streams and surface waters.

The pollution of surface waters by animal waste has led to a governmental permitting system with management controls. Under the Clean Water Act, federal regulations were developed in the 1970s under which concentrated animal feeding operations (CAFOs) were listed as point sources of pollution. These "CAFO regulations" help minimize water pollution to prevent environmental and health damages. No CAFO may discharge pollutants into surface waters unless authorized by a National Pollutant Discharge Elimination System (NPDES) permit. The permitting system operates to diminish the pollutants that enter the country's waters.

For pollutants entering surface waters, point sources are differentiated from nonpoint sources. Nonpoint sources of pollutants originating from forests and fields, construction sites, and other sources may enter waters as they are not subject to permitting requirements. Most farms are not CAFOs and so are not regulated as point sources. Pollutants from these farms are not effectively regulated under federal law. However, they are subject to lawsuits under trespass, negligence, and nuisance law.

Because a complete prohibition of any pollutants from entering surface waters point sources is infeasible, the Clean Water Act establishes a permitting system whereby persons apply for an NPDES permit. Under permits, permittees may have minimal qualifying discharges into surface waters that have a low risk of causing harm. The Act enables state agencies to take charge in administering the requirements for permits. In areas where a state is not authorized to issue permits, the regional Environmental Protection Agency (EPA) division performs this task.

For point source pollutants, state agencies adopt rules for discharge permits containing provisions that are similar to those in the federal regulations. Because discharges under a state permitting system cannot be less stringent than those required by federal law, a state cannot issue an NPDES permit in any situation where minimum federal effluent limitations have not been met. However, states can adopt more rigorous limitations to achieve water quality objectives. About 12,000 CAFOs are authorized by permits to engage in manure spreading activities that may result in legal discharges to surface waters.

Waste from other farms and from wild animals contain pathogens and nutrients that can create environmental and health problems. Most farms with livestock are not CAFOs, so their disposal of animal waste is not meaningfully regulated. It is unclear how much contamination of surface waters originates from these farms.

The pollutants

The top concerns with animal waste management are pathogens. We do not want to get sick, and manure often contains *Campylobacter*, *Escherichia coli*, *Listeria monocytogenes*, *Salmonella*, *Cryptosporidium parvum*, and *Giardia*. Some of these pathogens survive for months in animal manure. The pathogens may be transmitted to soils through the land application of manure, and then to surface waters. If vegetables or fruit come into contact with soil or water containing these pathogens, they can be transmitted to humans.

Fortunately, there is a rapid loss of viability of most viral, bacterial, and other pathogens. Moreover, management practices can minimize their movement into surface waters. But, occasionally, the contamination of leafy vegetables and other edible crops by manure leads to serious illness and death.

We are also concerned that the use of antibiotics at CAFOs is impairing water quality and creating environmental problems. The use of antibiotics at CAFOs contributes to the rise of antimicrobial resistance. Moreover, there is a strong relationship between on-site antibiotic use and resistance gene abundances in surface waters contaminated by animal manure.

Another concern is that animal waste discharges into wastewater lagoons may contain a variety of hormones. These include estrogens, natural and synthetic androgens and progesterones, and phytoestrogens associated with animal feed. Hormones detected in surface waters due to runoff from manure applications to fields may have adverse effects on aquatic environments.

Thus, the production of food animals accompanied by the disposal of their wastes present challenges. The CAFO regulations reduce the risks to enable food animals to be raised while safeguarding environmental quality and human health.

The regulation of CAFOs

The failure of the EPA to enforce the CAFO regulations in the 1980s led to a lawsuit resulting in an agreement to revise the federal regulations concerning NPDES permits. It took the EPA 11 years to adopt new CAFO regulations to safeguard water quality as required under federal law. Most states have adopted the federal requirements as state requirements.

The CAFO regulations only regulate manure from some CAFOs. Farms may be divided in four categories to determine whether the CAFO regulations apply to their manure disposal practices (Figure 18.1). The first category consists of farms that do not confine animals. These include all free-range cattle that are not confined for more than 45 days during a 12-month period. The second consists of confined animals at farms that do not meet the definition of a CAFO. Animal feeding operations with fewer than prescribed numbers of animals are not regulated by the CAFO regulations. The remaining two categories involve CAFOs and are distinguished by whether they need an NPDES permit.

CAFOs have long been regulated under a three-tiered system based on the existence of a discharge and their size, as determined by the number of animals

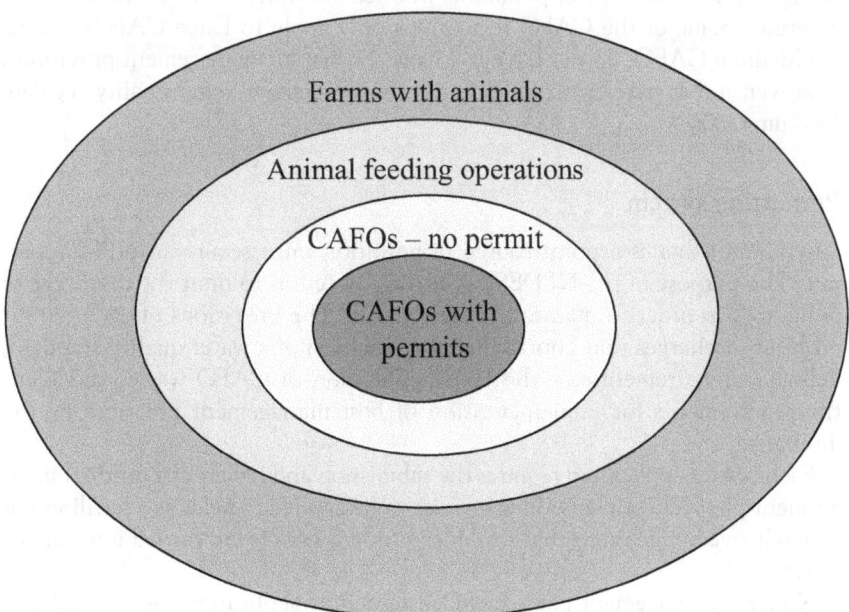

Figure 18.1 Four categories of farms, with one requiring NPDES permits

Table 18.1 Minimum animal numbers at a facility that constitute a CAFO.

Type of livestock	Large	Medium
Mature dairy cows	700	200
Cattle other than mature dairy cows or veal calves	1,000	300
Swine each weighing 55 pounds or more	2,500	750
Swine each weighing less than 55 pounds	10,000	3,000
Turkeys	55,000	16,500
Laying hens or broilers, if the operation uses a liquid manure handling system	30,000	9,000
Chickens (other than laying hens)	125,000	37,500
Laying hens, if the operation uses other than a liquid manure handling system	82,000	25,000

present at a facility. A discharge is the entry of pollutants into surface waters, which can only occur legally under a permit.

The three categories of CAFOs are Large, Medium, and Small. Large and Medium CAFOs are based on numbers of animals at a facility. For animals providing many of our food products, facilities with at least the numbers of animals listed below are Large or Medium CAFOs (Table 18.1). Farms with fewer numbers of confined animals that have a discharge of pollutants may be designated as Small CAFOs and would need an NPDES permit.

However, not every Large or Medium CAFO needs a permit. For some, if they do not have a discharge of pollutants into federal waters, they would not need a permit. Some of the CAFO regulations only apply to Large CAFOs, meaning Medium CAFOs do not have to follow the manure management provisions. However, a few states with heightened concerns about water quality regulate Medium CAFOs.

Permitting system

Any CAFO that is a point source of pollution must secure an NPDES permit. The purpose of the NPDES permitting system is to limit the discharge of pollutants in order to maintain water quality. The provisions of each permit prohibit discharges that contribute to a violation of a water quality standard, delineate requirements for the land application of CAFO waste, and identify requirements for implementation of best management practices by the discharger.

Each permit application requires the submission and review of a nutrient management plan. These plans allow manure to be applied to fields as a fertilizer but preclude overapplications that could lead to nutrients being carried into surface waters.

Nutrient management plans included in permit applications need to address four items. First, they need to identify site-specific conservation practices to be implemented. These would include buffers or equivalent practices to control runoff of pollutants to surface waters. Second, the plans should identify protocols for appropriate testing of manure, litter, process wastewater, and soil so that the producer can avoid applying excess nutrients.

Third, the producer needs to establish protocols for the land application of animal waste in accordance with site-specific nutrient management practices that ensure appropriate agricultural utilization of the nutrients. This is to avoid excess nutrients that could cause environmental problems. Finally, the producer needs to identify records that will be maintained to document the implementation and management of the waste. Under nutrient management plans, discharges that occur due to failure to follow the site-specific nutrient management practices violate federal law.

Attempts have been made to regulate proposed discharges from CAFOs. However, the EPA does not have the statutory authority for regulating proposed discharges (*National Pork Producers Council vs. EPA*, 2011). Conversely, a state's water quality program can incorporate more exacting controls to enable a state to regulate proposed discharges (Michigan Administrative Code, 2018).

Production and land application areas

The CAFO regulations apply to both a CAFO's physical areas of production as well as to land application areas that receive animal waste. Production areas include animal confinement areas, manure storage areas, raw material storage areas, and waste containment areas. Land application areas are lands under the control of an animal feeding operation owner or operator to which manure from a production area is or may be applied. Lands receiving manure from CAFO production areas are regulated because the applications involve the disposal of waste from a point source.

Production areas at Large CAFOs are not allowed to have a discharge, although rainfall event exceptions exist (US Code of Federal Regulations, 2018). If these regulations are enforced, CAFO production areas do not foul our surface waters. Rather, the impairment of waters by CAFOs occurs from accidental discharges and the land application of manure.

The EPA observed that runoff from land application of manure is a major route of pollutant discharges from CAFOs. Yet, the land application of manure is viewed favorably as a sustainable agronomic practice. It recycles nutrients for crop production and contributes to soil fertility. Manure applications at agronomic rates should not have a detrimental effect on water quality and so are allowed (EPA, 2003).

Manure needs to be "applied in accordance with site-specific nutrient management practices that ensure appropriate agricultural utilization of the nutrients" (US Code of Federal Regulations, 2018). By delineating how much manure can be applied, a nutrient management plan prevents unacceptable disposal practices that would impair water quality.

However, the overapplication of manure is not permitted because it can result in excessive amounts of nutrients accumulating in soils and running off into water bodies. Research shows a high correlation between phosphorus-saturated soils and impaired lakes, streams, and rivers (EPA, 2003).

In other situations, a producer may improperly manage manure applications, causing pollutants to leach into surface waters. This might involve applying manure on saturated soils or applications proximate to a rain event.

Agricultural stormwater discharges

The Clean Water Act defines a "point source" so that it does not include agricultural stormwater discharges and return flows from irrigated agriculture. Because the Act does not define agricultural stormwater, the EPA has delineated provisions in the CAFO regulations describing these discharges that apply to Large CAFOs.

Any precipitation-related discharge of manure from land areas under the control of a CAFO is an agricultural stormwater discharge if the manure was applied correctly under the CAFO's nutrient management plan (US Code of Federal Regulations, 2018). The application would ensure appropriate agricultural utilization of the nutrients. If a CAFO has a discharge that does not qualify as an agricultural stormwater discharge, it needs an NPDES permit prior to the discharge.

This definition means that the application of manure on fields by Large CAFOs may result in allowable agricultural stormwater discharges if the CAFO meets four components. First, the agricultural stormwater discharges need to occur as the result of a precipitation-related event. Any discharge not related to a precipitation-related event is not an agricultural stormwater discharge.

Second, a CAFO owner or operator has adopted specific conservation practices to control runoff from animal manure applied to land application areas. Third, the CAFO owner or operator has applied manure following site-specific nutrient management practices to ensure the appropriate agricultural utilization of the nutrients. Fourth, the CAFO has maintained records that document the implementation and management of its nutrient management practices.

If a CAFO owner or operator fails to adhere to documented nutrient management practices and there is a discharge, the discharge would be an impermissible point-source discharge. By adhering to the requirements of the nutrient management plan, a CAFO establishes qualification for agricultural stormwater discharges allowed under federal law. Without documentation, a CAFO owner cannot show that a discharge qualifies as an agricultural stormwater discharge. Thus, CAFOs with point-source discharges need NPDES permits, but CAFOs that only have agricultural stormwater discharges do not.

Public participation and review of plans

The Clean Water Act provides for public participation in the development and revision of effluent limitations to maintain water quality standards of surface waters. Since NPDES permits contain effluent limitations, some type of public participation is required prior to issuance of a permit (Centner, 2010). Furthermore, a permit application without a plan for managing nutrient pollutants does not allow meaningful public input into the development of effluent limitations as required by federal law.

An environmental group challenged the lack of participation in the approval process of CAFO NPDES permits in a lawsuit from Michigan (*Sierra Club Mackinac Chapter vs. Department of Environmental Quality*, 2008). Michigan's permitting program was found to be deficient because it did not provide for public participation as required by federal statutory requirements.

Public participation does not require a hearing. Rather, public participation requires an opportunity for input on required elements of nutrient management plans that are submitted to the permitting agency. Therefore, to provide public input, the required minimum elements of nutrient management plans must be available prior to issuance of a permit.

Moreover, when a permittee requests a modification to the terms of a permit, the public needs to be provided notice and an opportunity to be heard. Without an opportunity for public participation, citizens can file suit challenging a permitting authority's action. Whenever a modification to a permit was not validly adopted, the permittee is at risk for being liable for unauthorized discharges.

Another issue concerning the review of nutrient management plans is whether the permitting agency needs to review a submitted plan prior to issuance of a permit. The Clean Water Act requires the development of effluent limitations to control pollution. If a nutrient management plan is not reviewed, there is no way an agency knows whether the permittee has met the legal requirements for controlling pollution. An agency review is needed to prevent the approval of improper or inappropriate effluent limitations.

Lagoons and closure of facilities

The infrequent contamination of waters by lagoon collapses has spurred the regulation of these structures. Many states have enacted lagoon design provisions through legislation and agency regulations. The most common safeguards embody professional requirements for persons involved in designing manure storage structures and lagoons.

Common design specifications concern lagoon liners and capacity. States have rules prescribing liner requirements and the calculation of lagoon capacity determined by analyzing the volume expected to be generated over a designated number of days. For large lagoons and those with land application of liquid manure, governments may mandate the installation of groundwater monitoring wells.

Governments are also concerned about the proper disposal of manure and accompanying nutrients when lagoons or other manure storage facilities are closed. A state's closure rules may delineate a requirement for notifying officials when a facility is closed. In this manner, the state would be aware of the need for carrying out the closure plan. The state can then monitor closure procedures to ensure that animal wastes are disposed of properly.

Some state legislatures have adopted regulatory provisions delineating financial responsibility provisions for producers who go out of business. The closure regulations adopt provisions involving commercial or private insurance, guarantees, surety bonds, letters of credit, certificates of deposit, and designated savings accounts. By having operators place moneys in one or more of these instruments, the state has assurance that funds will be available to remedy problems that may occur if a lagoon is closed or an operation experiences financial difficulty.

Enforcement

The enforcement of the CAFO regulations involves an agency that addresses violations through enforcement mechanisms. A variety of mechanisms are used to oversee and respond to the problems accompanying the production of animals. They include warnings, civil penalties, injunctive relief, and criminal prosecutions.

In most cases, enforcement is administered by a state department of agriculture or the state environmental agency. There is considerable variation in the proficiency and rigor of enforcement efforts. States that support environmental quality may elect to place more resources and greater emphasis on enforcement matters. States with large populations may have more specialized staffs with greater expertise than exist in small states. These states tend to be more effective in enforcing the standards incorporated in various environmental regulations than others.

State agencies charged with enforcing regulations can only do so if their legislatures provide adequate funding. Due to political and fiscal reasons, some states do not allocate sufficient funds to employ the personnel and the equipment to engage in required enforcement activities.

Another concern is the willingness and competence of the enforcer to act against violations of regulations. Due to political or economic pressures, an agency or official may be hesitant to enforce pollution regulations. Agricultural interest groups exert considerable influence in some states. They may attempt to relax enforcement measures by political means.

The strict enforcement of regulations, however, is not always an optimal response. For some problems, an agency may issue a warning letter and then decline to do anything further. This may be appropriate because the agency is working with the operator in establishing meaningful measures to comply with the regulations. In other situations, the proof needed to establish the violation may be too formidable or the limited resources of an agricultural operator may recommend a warning rather than a fine or injunctive relief.

Given the hurdles that encumber effective enforcement, additional regulatory mechanisms may be recommended. Enforcement agencies may want to implement educational programs, document voluntary measures for adoption, and offer financial incentives to adopt pollution reduction technology.

Non-regulated animal production facilities

More than one million farms producing animals can dispose of their manure without meeting the requirements of the CAFO regulations. These farms are subject to nonpoint-source pollution requirements. Generally, nonpoint-source pollution provisions recommend voluntary practices known as best management practices that can be followed to minimize the potential adverse effects of an activity.

Best management practices are simply methods, measures, and practices that may be used to reduce or eliminate pollutants from waters. The US Department of Agriculture has recommended more than 40 different best management practices to meet conservation and stewardship goals, but these practices are generally voluntary, so not all farmers implement them.

Experiences with best management practices show that they may not solve nutrient pollution problems. However, they can intercept pollutants to avoid the impairment of waters. Due to costs and other reasons, satisfactory management practices are not being used on millions of acres of farmland receiving animal waste. Additional assistance may be needed to encourage greater use of practices that would prevent pollutants from livestock from entering surface waters.

A few states have imposed requirements beyond best management practices in an attempt to reach water quality goals. The most frequent requirements involve the preparation of nutrient management plans. These plans reduce nutrient discharges to minimize pollution of surface waters.

State and federal nonpoint-source pollution responses also involve financial and technical assistance. A range of planning, technical assistance, cost sharing, and public funding initiatives have been employed to assist with controlling nonpoint-source pollution. While such activities are positive and should be encouraged, they do not function as proscriptions against pollution.

Air pollution related to animal waste

Concern also exists about air pollutants from animal waste (National Research Council, 2002). The most troublesome air pollutants are ammonia, methane, nitrous oxide, hydrogen sulfide, and particulate matter. In some areas, aggregate emission goals may limit the numbers of animals that should be produced.

A combination of hydrolysis, mineralization, and volatilization convert nitrogen in animal waste to ammonia. Ammonia emissions from animal waste account for about 50 percent of the total ammonia emitted by all sources in the United States, and livestock rearing accounts for 64 percent of the world's anthropogenic ammonia emissions (Van der Heyden et al., 2015). Releases adversely impact atmospheric visibility, soil acidity, forest productivity, terrestrial ecosystem biodiversity, stream acidity, and coastal productivity (Galloway and Cowling, 2002). European producers in the Netherlands, Germany, and Denmark have adopted air scrubbers to reduce ammonia and other pollutant emissions from livestock facilities (Melse et al., 2009).

Direct emissions of ammonia and hydrogen sulfide from livestock and manure have been prominent and some feel this should be reported under federal law. In 2018, the Fair Agricultural Reporting Method Act specifically exempted air emissions from animal waste from the Comprehensive Environmental Response, Compensation, and Liability Act and the Emergency Planning and Community Right-to-Know-Act.

Livestock are major contributors of methane to the atmosphere. It is a greenhouse gas and contributes to global warming. Because it has a global warming potential 23 times that of carbon dioxide, it is particularly bad for the environment. The use of methane digesters can capture methane from lagoons and containment structures to use as a source of energy.

Hydrogen sulfide is particularly bothersome because of its odor. It may have negative localized effects on the environment. Some European producers are employing scrubbers and biotrickling filters to reduce odors that are offensive to neighbors (Oenema et al., 2012).

The movement of livestock and fans in animal housing facilities release particulate matter into the air that can have negative localized effects. Facilities raising poultry often release significant quantities of particulate matter. If particles are deposited in the lungs of humans, they can cause health problems.

Concluding thoughts

Governmental actions to preclude environmental degradation include laws and regulations that affect producers of animals. The Clean Water Act's provisions on water quality require the EPA to take action to prevent unauthorized discharges from CAFOs into navigable waters. Any production unit meeting the regulatory definition of a CAFO must comply with the requirements of the CAFO regulations. CAFOs with a discharge must secure an NPDES permit containing a nutrient management plan, unless the discharge qualifies as an agricultural stormwater discharge.

The Clean Water Act does not preclude the land application of manure. Rather, it delineates directives under which permittees and others should adopt best management practices to use nutrients for crop production and minimize pollution. While these provisions only apply to Large CAFOs, they enumerate a standard that defines acceptable application practices for all farms applying manure to fields and pastures.

American agriculture has been very successful in meeting numerous challenges over the past two centuries. Given problems with water pollution from animal waste, many farms with animals need to do more to adopt practices to prevent deterioration of environmental quality. Additional controls are needed to address situations involving the overapplication of animal waste and manure applications to saturated fields. These practices are contrary to good animal husbandry and should not be condoned.

Europeans have done more than the Americans in addressing problems from animal waste management. Yet, due to significant numbers of animals being raised in Europe, more regulatory controls may be implemented to control pollutants related to livestock production.

Foodwashing facts

1   Animal waste is an excellent source of nutrients to use for the production of crops.
2   Most farms dispose of their animal waste without oversight by environmental officials.
3   Large concentrated animal feeding operations need to have plans that minimize pollutant discharges.
4   Lapses in the enforcement of regulations governing the disposal of animal waste contribute to environmental and health problems.

References

Centner, T.J. 2010. Discerning public participation requirements under the U.S. Clean Water Act. *Water Resources Management* 24, 2113–2117.

EPA (Environmental Protection Agency). 2003. National pollutant discharge elimination system permit regulation and effluent limitations guidelines and standards for concentrated animal feeding operations (CAFOs): Final rule. *Federal Register* 68, 7176–7227.

Galloway, J.N., Cowling, E.B. 2002. Reactive nitrogen and the world: Two hundred years of change. *Ambio* 31, 64–71.

Melse, R.W., et al. 2009. Overview of European and Netherlands' regulations on airborne emissions from intensive livestock production with a focus on the application of air scrubbers. *Biosystems Engineering* 104, 289–298.

Michigan Administrative Code. 2018. Rule 323.2196.

National Pork Producers Council vs. EPA. 2011. 635 Fed. Rptr. 3d 738 (U.S. Court of Appeals for the Fifth Circuit).

National Research Council. 2002. *Air Emissions from Animal Feeding Operations, Current Knowledge, Future Needs. National Academies of Sciences, Engineering, Medicine.* Atlanta, GA: The National Academies Press.

Oenema, O., et al. 2012. *Emissions from Agriculture and Their Control Potentials.* TSAP Report #3, Version 2.1. International Institute for Applied Systems Analysis.

Sierra Club Mackinac Chapter vs. Department of Environmental Quality. 2008. 747 N.W.2d 32 (Michigan Court of Appeals).

US Code of Federal Regulations. 2018. Title 40, Sections 122.23, 412.31.

Van der Heyden, C., et al. 2015. Mitigating emissions from pig and poultry housing facilities through air scrubbers and biofilters: State-of-the-art and perspectives. *Biosystems Engineering* 134, 74–93.

19 Nuisances and product disparagement

Key questions to consider

1 How should nuisance law address activities needed for the production of livestock?
2 Should animal production facilities be able to expand in a manner that creates offensive odors for neighbors?
3 Are product disparagement laws needed?
4 Are new laws needed for criminalizing the release of unauthorized pictures, videos, and audio recordings?

The production of food products from animals is controversial. Smelly animal waste can interfere with the ability of neighbors to enjoy their properties. Crowded conditions at large production facilities can create filthy living conditions for the animals. Persons believing animals are being treated poorly may break the law to expose abusive practices, leading to monetary damages for farms and processing firms.

Agricultural groups and the meat industry are aware of public attitudes. They seek to preclude situations and information that detract from businesses involved in producing animals and marketing meat products. By sponsoring educational programs and supporting agritourism activities that help people learn more about the production of animals providing our food products, interest groups and businesses foster a positive image of food animal production.

Interest groups have been proponents of three groups of legislative ideas to offer advantages for agricultural production and processing facilities (Table 19.1). These ideas have been adopted by states legislatures as anti-nuisance, product disparagement, and anti-investigation laws.

In the 1960s, agricultural interest groups were concerned about nuisance lawsuits. The result was anti-nuisance legislation to protect agricultural operations from nuisance lawsuits. In the 1990s, the issue became the disparagement of food products. Twenty years later, the animal industry sought expanded protection through new state legislation aimed at penalizing unauthorized investigation and publication of conditions at animal production and processing facilities. These

Table 19.1 Three groups of state legislation to offer advantages to agricultural producers

Anti-nuisance (right-to-farm) laws	To preclude persons from using nuisance law to stop production activities
Product disparagement laws	To create liability for false statements that lead to economic losses
Anti-investigation (ag-gag) laws	To punish persons from gaining employment through lies and disseminating unauthorized pictures, videos, and audio recordings

anti-investigation laws became known as the "ag-gag" laws. Ag-gag laws address situations that can adversely affect the profitability of animal production and marketing firms.

Nuisance law

Each state's common law includes principles of public and private nuisance law. Private nuisance law allows people to stop activities that are objectionable for the location and interfere with surrounding property owners' use and enjoyment of their properties. For an activity that adversely impacts the order and economies of the public at large, an action in public nuisance can be used to stop the activity. A public nuisance typically arises on a defendant's land and interferes with a public right.

When some people move next to farms, they feel their ownership of property allows them to take action to stop nearby objectionable activities. Because nuisance law involves community standards, the majority can object to an activity and use nuisance law to stop it. Even when farmers are engaged in established agricultural activities that are appropriate for the countryside, neighbors can use nuisance law to stop the activity.

A nuisance cause of action is important because neighbors objecting to an agricultural activity can secure an injunction. Through a court order, the neighbors can stop an offensive activity so that it will no longer be offensive to the community. Most farmers are aware that their activities can be found to be annoying to others. Smells from animal operations are the most obvious. Dust created by cultivation practices, noise from harvesting crops after dark, and vermin from a farm are other examples of conditions that lead neighbors to complain.

Court orders that end farming activities are economically detrimental to agriculture. In many cases, stopping activities increases the costs of producing agricultural products. In some cases, buildings and machinery can no longer be used, so farmers suffer financially. For a few farmers, the inability to engage in an activity will force them to go out of business.

Yet, living next to a large animal feeding operation can be unhealthy. Concentrations of animals can be especially unfair to neighbors as they tend to be smelly and the disposal of manure leads to objectionable odors. Discharges of hydrogen sulfide and ammonia from animal facilities can adversely affect neighbors' health (Donham et al., 2006). Should neighbors be able to bring a nuisance lawsuit for the smells, flies, and rodents accompanying animal production that denigrate their health?

Anti-nuisance laws

The unfair situation where neighbors used nuisance law to enjoin longstanding farm practices led agricultural interest groups to advance anti-nuisance legislation. Under the laws, agricultural operations can qualify for immunity from nuisances and a court cannot order an activity to be stopped. These anti-nuisance laws are known as "right-to-farm laws" because they delineate rights for farmers to continue with their farming activities.

Every state has passed an anti-nuisance law to preclude some nuisance lawsuits. They embrace the concept that persons living in rural areas near agricultural operations accept the annoying activities as part of their location choice. Anti-nuisance laws say that existing land uses may continue even when changes in the area cause the facility to be objectionable to the neighbors. Legislators have made a policy decision to support existing economic activities by curtailing the ability of neighbors to enjoin activities and practices. This supports the retention of farmland.

Anti-nuisance laws thereby shift the balance of competing property rights so that some annoying agricultural activities are condoned. In other situations, neighbors are able to employ nuisance law to stop objectionable practices and anticipated nuisances.

Anti-nuisance laws are especially important for farms with livestock. Raising animals involves manure, and manure is smelly. Americans are generally unfamiliar with animal production and do not appreciate that the use of animal manure as a fertilizer on nearby cropland is a good husbandry practice. Anti-nuisance laws have given many agricultural activities sufficient protection from nuisance lawsuits so that farmers can continue with their operations.

As farm groups realized the power of these laws, they were amended to provide even greater protection to agriculture and associated industries. Today, the state laws are quite different in the protection they offer agricultural producers. While most provide reasonable protection for existing farming activities, a few delineate favoritism that allows new nuisances to adversely affect neighbors. The delineation of three doctrines highlights major distinctions of anti-nuisance laws (Figure 19.1).

Laws that delineate a coming-to-the-nuisance principle embody the original intent to help existing farmers. A second group of laws imposes time limits for commencing nuisance lawsuits so nuisances are resolved within a reasonable period of time. Anti-nuisance laws allowing new technologies and major expansion enable persons to adversely affect neighbors.

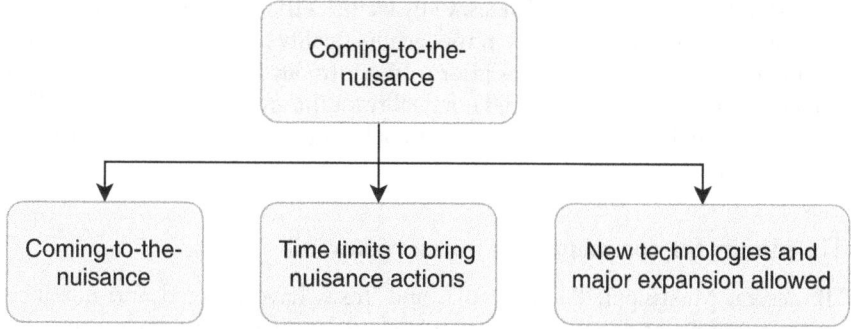

Figure 19.1 Major anti-nuisance doctrines

Coming-to-the-nuisance

The original intent of the anti-nuisance laws was to protect viable agricultural operations from new neighbors who moved to the country. These laws incorporated a "coming-to-the-nuisance" doctrine under which farmers can continue with bothersome activities in situations where complaining neighbors moved into the area. Persons who move next to a smelly or dusty agricultural operation should accept the annoying activities as part of their choice to live in the country.

A typical scenario involves a residential subdivision being built near existing farms. After a few years, the new neighbors object to a farming activity and seek to stop it under nuisance law. Due to the provisions of an anti-nuisance law, neighbors moving to the nuisance cannot use nuisance law to stop a qualifying agricultural activity.

Anti-nuisance laws incorporating the coming-to-the-nuisance laws have adopted a policy that grants rights based upon priorities in land usage. Neighbors are precluded from employing nuisance law to stop preexisting activities. Activities that become a nuisance due to changes in neighboring land uses receive protection by an anti-nuisance law. Thereby, activities that were objectionable when commenced and operations generating new nuisances do not receive protection against nuisance lawsuits.

Some farmers have not appreciated these qualifications. In a Georgia case, a farmer constructed 26 chicken layer houses near existing residential neighbors. More than 500,000 chickens were soon creating problems for the neighbors, and the neighbors filed a nuisance lawsuit against the farm to eliminate the flies and offensive odors generated by the chickens. The neighbors asked the court to enjoin the farm from further business activity.

The Supreme Court of Georgia examined the new anti-nuisance law and observed that its protection did not apply (*Herrin vs. Opatut*, 1981). The anti-nuisance protection was limited to situations where nonagricultural land uses extended into agricultural areas. Only facilities that have become nuisances as a result of changed conditions in the locality qualify for the defense. Since the neighbors resided in their homes prior to the introduction of the chickens, the farmer with the chicken houses did not qualify for the anti-nuisance defense. The chickens created a nuisance so the court could order the farmer to take actions to end the nuisance.

Time limits for commencing a nuisance lawsuit

Minnesota, Mississippi, Pennsylvania, and Texas have adopted anti-nuisance laws containing provisions on time periods to extinguish nuisance causes of action. The laws basically provide that once an agricultural operation has lawfully been in operation for a stated period of time, no nuisance cause of action exists. However, the operation must have existed substantially unchanged since the established date of operation.

In addressing the meaning of the Texas anti-nuisance law, a court found that two conditions were required to meet the requirements of the law barring a nuisance lawsuit (*Ehler vs. LVDVD, L.C.*, 2010). First, the agricultural operation needed to have been in business lawfully for more than one year before the nuisance action was filed. Second, the operation needed to show that the conditions and circumstances complained of as constituting the basis for the nuisance action had existed substantially unchanged. Since the operation met these conditions, it had protection against the plaintiffs' nuisance action.

Pennsylvania's anti-nuisance law incorporates an idea whereby operations with approved nutrient and odor management plans are afforded protection against nuisance lawsuits (Pennsylvania Statutes, 2017; Pennsylvania Code, 2018). Nutrient management plans are planning documents for determining appropriate practices to manage manure at animal feeding operations. Odor management plans are plans identifying the practices, technologies, standards, and strategies to be implemented to manage the impact of odors generated from animal housing or manure management facilities. An operation is accorded the anti-nuisance protection if it has lawfully been in operation for one year or more prior to the date of the plaintiff's lawsuit.

Via nutrient management plans, agricultural operations in Pennsylvania that physically expand their facilities can continue to have a defense against nuisance actions so long as their activities have been addressed by the plan. Concentrated animal operations that expand can develop and implement an odor management plan to qualify for protection against nuisance lawsuits.

Expanding operations and technologies

Livestock facilities that expand or adopt new technology and create a nuisance present a difficult issue. If an animal production facility qualifies for anti-nuisance protection with respect to existing neighbors, does its expansion also qualify?

Similarly, should the adoption of new technology be protected due to the earlier presence of the animal production facility?

In some states, agricultural interest groups convinced state governments that agricultural producers deserved protection for expanded operations and new technologies. They deleted the coming-to-the-nuisance provision and wrote laws declaring that agricultural production was more important than other land uses. The justification of these one-sided laws was that people need food and the production of food is beneficial to a state's economy. Given these conclusions, agricultural groups felt that no one should interfere with existing and new agricultural practices, even when they are objectionable.

However, nuisance law is an equitable remedy that was developed over centuries and was intended to reconcile clashing property rights. Courts could fashion equitable remedies that were fair to the competing interests. A one-sided anti-nuisance law that places agriculture above other property uses removes fairness. The equitable remedy of nuisance is replaced by a law that enables farmers to engage in activities that severely harm neighboring property owners.

A one-sided anti-nuisance law stinks. Residential neighbors in rural areas should enjoy the right to a healthy environment. Neighbors deserve to be able to use their properties for reasonable activities. A one-sided anti-nuisance law favoring agricultural producers does not balance the rights of farmers with the rights of neighbors. Instead, it completely overrules the rights of neighboring property owners. Such a law is inimical to public health and community well-being.

An example of a one-sided anti-nuisance law is highlighted by a lawsuit from Indiana (*Dalzell vs. Country View Family Farms, LLC*, 2013). A farm that had been growing corn and soybeans for more than 50 years was converted to a pig farm. Suddenly, the neighbors were living next to 2,800 hogs. The neighbors objected to the odors and filed a nuisance suit. In analyzing Indiana's anti-nuisance law, the court found that an operation can convert from one type of operation to another, such as crop farming to a pig farm.

The one-sided Indiana anti-nuisance law says neighbors must accept the obnoxious conditions accompanying 2,800 hogs moving next door. The law precluded any relief for the Dalzells from the awful stench of 2,800 pigs located within feet of their property. Because the Indiana anti-nuisance law allows properties to change from one type of agricultural operation to another, starting a new pig operation on crop land was protected by the law. If the Dalzells do not want to live next to the new pig farm, they should move.

But, why should the Dalzells have to move to avoid the smells from the new pig farm? The facility should not have been built next to a residential property. The protection of new nuisances allows situations that are unfair to neighboring property owners. The laws generate hardship. Anti-nuisance laws that cause too large an interference with the rights of neighboring property owners may violate constitutional provisions (Centner, 2006).

Today's rural America includes agricultural, recreational, and residential land uses. Property owners have obligations to neighbors, and our laws should strive to reconcile competing rights in an equitable fashion. While many state anti-nuisance laws provide reasonable protection to existing animal production facilities, a few allow extremely offensive conditions that adversely affect neighbors.

Product disparagement laws

In 1989, CBS aired a *60 Minutes* episode about a plant growth regulator called Alar, a pesticide used in food production (Negowetti, 2015). The episode reported that Alar was being used on Red Delicious apples, and experiments with feeding animals Alar suggested it was a potential carcinogen. The report inferred that eating apple products might pose a risk of cancer.

The public response to this Alar news report was devastating to fresh Red Delicious apple sales. Prices dropped, fresh supplies were converted into processed products, and Washington State apple growers lost about $130 million.

The Washington State apple growers sued CBS, alleging product disparagement. However, the lawsuit was dismissed as there was insufficient evidence to prove that the facts in the news report were false (*Auvil vs. CBS "60 MINUTES"*). Scientific research had indicated that Alar breaks down into a carcinogen. Thus, there was no disparagement under common law.

The damages inflicted on apple growers led agribusiness and food interest groups to advocate new legislation. The legislative ideas spread to state legislatures, and several adopted product disparagement laws in the 1990s. Some people referred to these laws as the "veggie libel laws" (Semple, 1995). The laws establish statutory torts under which a false statement can lead to liability for resulting economic injuries.

Under the product disparagement laws, persons who make false accusations that adversely affect the market of an agricultural product can be held liable for damages. However, proving that a statement is false tends to be difficult. For example, if a statement is made that a particular pesticide may cause cancer, this is an opinion and so is not false. Yet, such a statement could lead to an adverse public response. Thus, the veggie libel laws are not very effective.

Another problem with these laws is they tend to offend constitutional free speech. Individuals, consumers, and researchers have a constitutional right to raise legitimate questions about food safety and quality.

Anti-investigation (ag-gag) laws

State legislatures have been concerned about persons who enter agricultural facilities without permission to take pictures, videos, and recordings for nearly 30 years. A few states enacted laws in the 1990s to prohibit entry to animal facilities with the intent to commit criminal defamation (Shea, 2015). Others enacted animal terrorism laws targeting vandals and activists who release laboratory animals. Congress was concerned about animal enterprise terrorism and enacted the Animal Enterprise Protection Act in 1992.

In the mid-2000s, members of animal rights groups and others sought employment at facilities involved with livestock production and marketing in order to expose lousy conditions and abusive practices. Using lies on employment applications, they gained entry, gathered pictures, videos, and audio recordings, and then released the unauthorized materials to the public.

One of the more publicized incidents involved an undercover investigation at a slaughterhouse in California by the Humane Society of the United States. The video recording showed shocking animal abuse by workers attempting to force ill animals to walk to slaughter (Pitts, 2012). Moreover, the meat from these animals was being processed and sold to schools for school lunch programs. The state responded with a law requiring euthanasia for nonambulatory animals (California Penal Code, 2017).

Subsequently, undercover investigators at other facilities gained pictures and videos of filthy conditions and animal abuse. Dead chickens in laying houses and maggots in manure piles showed deplorable conditions. Publication of the undercover videos of conditions and situations at animal production facilities were disastrous for the industry. They led to product recalls, boycotts, the termination of purchasing contracts, and bankruptcy (Shea, 2015).

The financial losses and bad publicity of these investigations led agricultural interest groups to request new laws to punish persons illegally gathering information. In many of the cases, the persons had gained entry through lies on their employment applications. In other situations, persons violated employment contracts that prohibited the employee from recording images and sounds. Agricultural interest groups proposed new laws with penalties for persons involved in fraud or releasing unauthorized information.

Legislatures referred to these legislative proposals as interfering with agricultural production. The laws made it illegal to photograph, video, or record sounds at an agricultural facility without the consent of the property's owner, and also to distribute such footage without consent (Shea, 2015). Yet, existing laws on conversion, trespass, and fraud already provided remedies for such interferences. Rather, it was apparent that the laws were intended to stifle investigative reporting. The laws quickly became known as "ag-gag" laws.

A majority of the newspaper editorials were against the ag-gag laws because they would curb investigative reporting. It was also noted that the laws would increase distrust in our food supplies. If someone is engaged in a practice that abuses animals or violates safety rules, the public has a right to learn about the activities.

Groups opposed to ag-gag laws maintained a simple persuasive message: why ban cameras if there is nothing to hide? The laws are not needed if producers and businesses are conducting their businesses legally and are not engaged in abusive or unethical practices. Even rural newspapers noted that farms engaged in proper activities do not need these laws (Landfried, 2013).

While powerful agricultural interest groups advanced ag-gag laws in dozens of states, the laws were adopted in only about a dozen. This was due to animal welfare activists winning the public opinion battle. With the colloquial name "ag-gag," the public found the anti-investigative provisions to be contrary to free speech. People wanted to know about persons and businesses that were engaged in animal abuse or other objectionable conditions. Given public opposition, state legislators and governors backed away from supporting the legislation.

Restrictions on investigative reporting

In their efforts to prevent the fraudulent release of information from agricultural production facilities, several ideas were proposed for ag-gag laws (Figure 19.2). First, gaining entry to a facility by false presences is illegal. This would involve a false statement by a person in order to enter the facility.

Second, gaining employment by false presences is illegal. This would involve a lie on an employment application or an employment contract. Third, a person who obtains documents of an agricultural production facility by force, threat, misrepresentation, or trespass is interfering with the facility. Such coercive or fraudulent conduct is a crime.

Fourth, a person who records an image or sound from the facility without permission is committing an illegal interference. This may involve taking something without permission or a violation of an employment contract.

It was also suggested that persons with evidence of animal abuse should be required to turn it over to the police within 24 hours or other short time period (Shea, 2015). This idea has not been adopted because it would force undercover investigators to blow their cover within a day or two of recording the first evidence of abuse. This would prevent evidence of a pattern of abuse at a given facility.

Challenges to ag-gag laws

An ag-gag law will not be effective unless alleged violators are prosecuted. With the adoption of the Utah ag-gag law, the state decided to prosecute Amy Meyer for taking a video of a sick cow being moved by heavy equipment outside a slaughterhouse. The video was filmed while Ms. Meyer was standing on public property and showed animal abuse. But she had interfered with an agricultural facility by releasing the video.

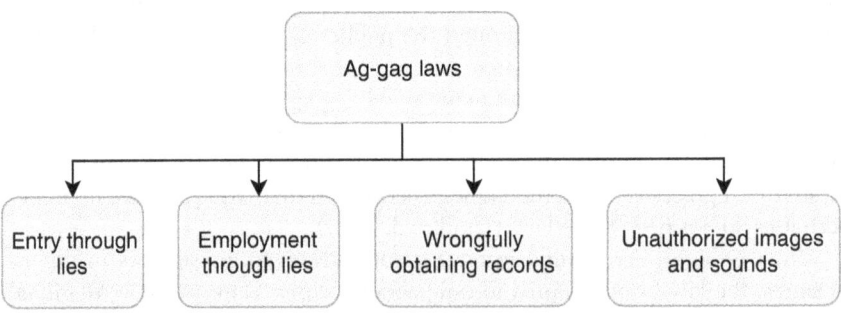

Figure 19.2 Distinct provisions of ag-gag laws making conduct illegal

While Ms. Meyer's video drew attention, the state's lawsuit made national news. The public's reaction was overwhelmingly negative for the state and the animal industry. The state finally decided to drop the lawsuit. But a significant point had registered with the public. Ag-gag laws were attempts to hide animal abuse and bad practices. Legislators supporting ag-gag laws were condoning animal cruelty.

Subsequently, the Animal Legal Defense Fund and other groups sued Utah state officials, challenging the Utah ag-gag law as an unconstitutional restriction on free speech (*Animal Legal Defense Fund vs. Herbert*, 2017). The state argued that criminal activity should not be accorded First Amendment speech protection. However, the court noted that some of the lies criminalized by the Utah ag-gag law enjoy First Amendment protection. The court found that the law was unconstitutional in suppressing broad swaths of protected speech in violation of the First Amendment.

A second law that was challenged was the Idaho ag-gag law (*Animal Legal Defense Fund vs. Wasden*, 2018). Again, the Animal Legal Defense Fund and other groups had challenged the ag-gag law as an unconstitutional limitation on free speech. It was also argued that the ag-gag law violated the Fourteenth Amendment because it effected discrimination motivated by animus toward animal welfare groups. The district court found no rational basis for the law as existing laws against trespass, conversion, and fraud already provided avenues for protecting private properties.

On appeal to the federal circuit court of appeals, two of the law's provisions were found to be offensive. The court concluded that the criminalization of misrepresentations to enter a facility offended free speech. The court also found that the provisions banning audio and video recordings covered protected speech rights protected by the First Amendment.

The circuit court's ruling meant that the state could criminalize misrepresentations used to obtain records and secure employment. The court recognized that these provisions of the Idaho ag-gag law were permissible.

Debates will continue on the merits of ag-gag laws. While the industry has reason to be concerned about fraudulent employment applications, unauthorized investigations, and violations of employment contracts, enforcement of these problems is possible under existing tort law. Therefore, legislatures may decide that ag-gag laws are unnecessary.

Foodwashing facts

1 Neighbors who stop objectionable agricultural activities under nuisance law can adversely affect agricultural producers.
2 Anti-nuisance laws rebalance nuisance rights to allow farmers to continue with agricultural activities.
3 Persons falsely disparaging food products are violating the law.
4 Curbing investigative reporting can interfere with the freedom of speech.

References

Animal Legal Defense Fund vs. Herbert. 2017. 263 F. Supp. 3d 1193 (U.S. District Court, Utah).

Animal Legal Defense Fund vs. Wasden. 2018. 878 F.3d 1184 (U.S. Ninth Circuit Court of Appeals, Seattle).

Auvil vs. CBS "60 MINUTES." 1995. 67 F.3d 816 (Ninth Circuit Court of Appeals, Seattle).

California Penal Code. 2017. Section 599f.

Centner, T.J. 2006. Governmental and unconstitutional takings: When do right-to-farm laws go too far? *Boston College Environmental Affairs Law Review* 33, 87–148.

Dalzell vs. Country View Family Farms, LLC. 2013. 517 Fed. Appx. 518 (Seventh Circuit Court of Appeals, Chicago).

Donham, K.K., et al. 2006. Assessment of air quality at neighbor residences in the vicinity of swine production facilities. *Journal of Agromedicine* 11(3/4), 15–24.

Ehler vs. LVDVD, L.C. 2010. 319 S.W.3d 817 (Texas Court of Appeals).

Herrin vs. Opatut. 1981. 281 S.E.2d 575 (Georgia Supreme Court).

Landfried, J. 2013. Bound and gagged: Potential first amendment challenges to "ag-gag" laws. *Duke Environmental Law & Policy Forum* 23, 377–402.

Negowetti, N.E. 2015. Opening the barnyard door: Transparency and the resurgence of ag-gag & veggie libel laws. *Seattle University Law Review* 38, 1345–1398.

Pennsylvania Code. 2018. Title 25, Section 83.741.

Pennsylvania Statutes. 2017. Title 3, Section 954.

Pitts, J. 2012. "Ag-gag" legislation and public choice theory: Maintaining a diffuse public by limiting information. *American Journal of Criminal Law* 40, 95–110.

Semple, M.W. 1995. Veggie libel meets free speech: A constitutional analysis of agricultural disparagement laws. *Virginia Environmental Law Journal* 15, 403–442.

Shea, M. 2015. Punishing animals rights activists for animal abuse: Rapid reporting and the new wave of ag-gag laws. *Columbia Journal of Law & Social Problems* 48, 337–371.

20 Conclusions

Key questions to consider

1   What is being done to guarantee that products from animals are safe to eat?
2   Should we interfere with existing animal production practices?
3   Are producers using too many objectionable inputs?
4   What suggestions might we offer to enhance animal production practices?

Consumers in the United States enjoy a lot of choices in selecting food to eat. In fact, some feel there are too many choices and are overwhelmed by so much information. The media bombards us with advertisements for tasty food, recommendations for healthy food, and reports of unsafe food. Labels at stores and on products contain information on amounts of ingredients, health facts, and the absence of ingredients. Sometimes we ignore most of the information. In other cases, we have difficulty in comprehending it. When we buy food products, how much time and effort do we really want to spend on each item we select to eat?

Meat and animal products (jointly referred to as animal products) present their own challenges. Having safe products is our top priority. Many meat and dairy products are perishable, and a lapse in refrigeration or waiting too long to consume products can result in illness or even death. For some products, we want to know about the production practices and inputs that were administered to the animals at their production facility. In other cases, we want information on social issues so we can avoid products from animals produced with objectionable practices.

We need to recognize that the production of animals and marketing their products involve business decisions. Animal producers want to make a profit. To reduce costs, producers adopt practices and procedures that affect the well-being of food animals and the safety of food products. Consumers can object to facilities and inputs as they make selections of animal products. Some consumers are willing to pay more to secure a product that is free from objectionable production practices and inputs. The ability to purchase these products depends on information affixed to the label or conveyed at the point of sale.

The chapters of the book were grouped in four categories: safety, production facilities, inputs, and social issues. In describing the topics, both positive and negative aspects about animal production and their products were discussed and viewpoints were presented. Each chapter concluded with a few factual statements that highlight major issues.

With this information, what might be done to adjust production and marketing practices so that the industry better serves the consuming public? In this chapter, short summaries of the four major categories of topics highlight the issues. Subsequent suggestions offer ideas for making adjustments in the production, marketing, and regulation of animal production and products. With these suggestions, consumers and the industry can initiate changes that address undesirable features. This includes enabling consumers to voice opinions and exercise rights that are keystone features of a democracy.

Safety and the food animal industry

Our ability to have safe food for consumption is dependent on the actions of others. We need to appreciate the many efforts that are being made every day by producers, slaughterhouses, manufacturing facilities, marketeers, and retailers to keep our meat and animal products safe. There are so many opportunities for something to occur that could compromise the quality or safety of a product. How do our production and marketing operations function to maintain such a safe marketplace?

Let's first look at some potential safety pitfalls. During production, an animal may consume a substance that causes its products to be unsafe for consumption. At the slaughterhouse, a meat product may be contaminated with feces containing deadly bacteria. A manufacturing facility may fail to adequately clean a piece of equipment, enabling a pathogen to contaminate a product. A shipper may experience problems with refrigeration, causing products to be unsafe. A storage unit or retail facility may have rodent or insect pests that lessen the quality of a product. These are just a few examples of situations that need to be avoided in order to have safe products.

Despite the many opportunities for compromising the safety of our food, an overwhelming majority of our food products are safe. Due to laws assigning responsibility and liability, producers and marketers have incentives to keep food products safe. In addition, we have governmental and industry safety checks and balances to preclude situations and conditions that could lead to unsafe animal products.

Federal laws, regulations, and inspections by the US Department of Agriculture (USDA) and the Food and Drug Administration (FDA) govern the production and marketing of animal products. Both of these agencies oversee the application and enforcement of regulatory provisions designed to keep food products wholesome. States have also undertaken efforts to keep our food safe. Thus, the safety of food products is grounded on procedures that reduce risks of contamination and liability for products causing damages.

Raising food animals requires knowledge about their needs and practices available to encourage weight gain and good health. Although humans have been raising animals for thousands of years, practices and technologies have changed.

Today's producers of food animals and marketers of animal products are businesses that exist because they are profitable. Producers with high production costs are forced out of business. Marketing firms that cannot make profits cease to exist.

This competitive marketplace extends to processing and manufacturing facilities for animal products. The industry has adopted technologies and practices to reduce costs while maintaining safety standards. New facilities with labor-saving equipment and the capacity to handle large volumes have lower costs than existing facilities, causing the latter to go out of business.

The industry has also responded to consumer preferences. There are market niches for specialized animal products, and a wide variety of meat and animal products are available for consumers who want certain attributes. We can learn about products' attributes from labeling information and avoid products that we feel are unhealthy or have a connection to an input or practice we do not like.

It is challenging to develop and maintain markets for specialized products and to convey information and guarantees about the attributes of these products. There must be regulatory controls and oversight to curtail bad practices, adulterated products, fraud, and dishonesty. The legal system needs to impose penalties for violations. Given these facts, what might be adjusted to enhance consumer safety while facilitating a viable industry that is able to produce and market a wide variety of animal products?

Suggestions for safety and the industry

Three suggestions may be offered to further augment the safety of our food supplies. Two are related to the enforcement of mislabeling regulations and the third recommends continued support for innovative technologies.

Allow states to enact safety regulations

The Poultry Products Inspection Act and the Federal Meat Inspection Act only allow the USDA to regulate mislabeling and fraud involving meat products. Both Acts preempt additional requirements by state governments. This means that a state cannot enact a mislabeling law for prosecuting claims of mislabeled meat products.

Due to the lack of resources and other reasons, the USDA is not able to fully enforce its regulations governing mislabeled meat and dairy products. This suggests that the federal laws should be amended so that states would have authority to combat fraud and mislabeling. If we are serious about helping consumers, we need to reinforce federal law and allow others to help the USDA fulfill its mission of protecting consumers from fraudulently labeled products.

Augment enforcement

Given the limited resources and personnel of governments, a new legal structure may be advantageous that would facilitate additional persons to help oversee food safety and fraudulent labeling practices. One idea is to adopt a "citizen suit" provision modeled after the federal citizen-suit provisions of environmental legislation. Such a provision could enable non-governmental groups to take actions in situations where violations have occurred and the federal government has not acted.

Encourage and support technology

New technologies enable us to reduce costs of producing and marketing animal products. Technologies also enable us to guarantee the safety of products, monitor conditions related to animal health, and to use more sustainable practices. Although we do not want to compromise safety, we should continue to incorporate new technologies into production and marketing processes. To help the public become more receptive to new technologies, the industry needs to do a better job of explaining their merits.

Potential problems at animal production facilities

During the past 20 years, the practices involved with the production of food animals have changed. New technologies make it profitable to raise many animals at a single farm. The world's increasing population means we need more food. Rising incomes in developing countries have led to increased consumer demand for meat products. The depletion of wild fish stocks has led to more farm-raised seafood products to meet consumer demands.

Meanwhile, our perspectives on the treatment of food animals have changed. We object to their suffering and criticize crowded production conditions and cages. Our sentiments have led to changes at production facilities and slaughterhouses. While these changes have been positive, some people are advancing additional proposals to prohibit practices and limit producers in the technologies they can use. A majority of these limitations would increase production costs. In turn, they would adversely affect consumers who are food insecure.

One of the most controversial technologies has been the production of large numbers of animals at individual farms. Some have labeled this production as factory farming. Yet, this term fails to consider the fact that animals are not manufactured and that 90 percent of our farms are family farms (MacDonald and Hoppe, 2017). The animals are living creatures being raised for food with different technologies than were available in the past. Labeling a farm as a factory due to the effective use of new technologies incorrectly describes the changes. New production practices and inputs assist in raising animals, not manufacturing products.

Looking further at the technologies being employed at farms, many augment safety and help maintain favorable living conditions for the animals. This is especially true for the technologies and practices being used at modern farms. Video and audio technologies can monitor the health of animals to facilitate timely veterinary interventions. This helps prevent illness and animal suffering. Large facilities are more likely to have nutrient management plans for handling animal waste in a sustainable manner. They also tend to have more detailed records so that less feed and energy are consumed in the production of our animal products.

Because animals are living creatures, we are concerned about their treatment while they are raised for food. However, treatment depends on the producer, not the size of the farm. Most producers are conscious of the conditions and treatment of their animals. Abused, stressed, and sick animals are not productive. They do not gain weight as fast, and the quality of their products may be inferior, meaning that buyers pay less for these animals. Because profitability is related to weight gain and quality products, producers want contented, healthy animals. Farms that condone conditions that are unhealthy or that mistreat animals are not going to be as profitable.

Consumer activism has been important in improving the living conditions of many food animals. Because of consumer pressure, producers are giving animals more freedom in expressing themselves and more space. Regulatory oversight and oversight by marketers have curtailed abusive and egregious practices. Simultaneously, conditions at some production facilities, both large and small, are less than ideal. The competitive market for meat products means producers are under pressure to reduce costs, and cost-saving practices can adversely affect animal well-being. We need to maintain efforts that monitor the well-being of animals being raised for food.

Suggestions for production facilities

Three ideas are offered concerning the production of food animals. Since most producers of our animals maintain good conditions due to the fact that animals gain weight faster if they are comfortable, and marketers of animal products have adopted polities that have reduced bad practices, our main concerns should focus on seafood products and voluntary marketing labeling of products.

Preserve wild fish stocks

We need to encourage our governments to do more to protect wild fish stocks and prevent overfishing. Depleted stocks have reduced the amounts of wild fish that can be harvested, leading to the need for greater production at fish farms. Through better management of fisheries, we can increase the amounts of wild fish available for consumption.

Reduce sources of mercury pollution

It is well known that mercury can accumulate in fish to levels that are unhealthy for persons consuming the products. Fishermen and persons consuming seafood products need help in keeping mercury pollution away from seafood products that are consumed by other animals and people. Greater regulatory efforts are needed to reduce sources of mercury pollution in surface waters.

Use marketing information to encourage production practices

Numerous ideas have been advanced to enact regulations to control practices at production facilities related to inputs, husbandry practices, and animal well-being. Many of the proposals would impose costs on producers, slacken innovation, and detract from free enterprise. Governments already have addressed the most egregious problems. Allow the marketplace to use labeling information to allow consumers to voice their preferences for production and marketing practices. Labeling information on cage-free eggs and antibiotics shows that consumers can effectively alter production practices. Voluntary labeling allows producers to be innovative in adjusting and adopting technologies at their facilities.

Consumer information on inputs

Consumer groups have been active in learning more about the inputs being employed in producing food animals. They have also formed opinions about these inputs and have sought to reduce or even prohibit their use. Overall, we should be pleased with these consumer efforts, as they have contributed to changes that are beneficial to our health and the well-being of food animals.

All of us want wholesome food products and worry about the use of inputs that may compromise safety. However, we need to realize that defining wholesome is difficult. Because we cannot examine the effects of traces of pesticide or beta agonist residues in food products on humans directly, we rely on animal studies. The observations of these studies are difficult to interpret and relate to humans. Moreover, we need to realize that most foods we eat contain traces of substances that are unhealthy. For example, most public drinking water supplies contain traces of arsenic, yet these supplies are deemed safe and we continue to drink water from these sources.

In approving the use of hormones, beta agonists, and pesticides, our federal government has conducted extensive studies of the potential for adverse health effects on humans. Given the findings of these scientific studies, the government established limitations that allow these inputs to be used due to their benefits. Yet, there are different opinions about their safety, and a few people may reach a different conclusion. While minority viewpoints need to be heard and respected, we should adopt policies on food safety that are based on scientific viewpoints held by the majority of scientists.

Antibiotics are used in the production of food animals to enhance weight gain, keep animals healthy, and to prevent animals from suffering. The overuse of antibiotics in animal production is believed to be negatively related to the development of antibiotic-resistant bacteria. Given our need to slacken antibiotic resistance to preserve the efficacy of antibiotics for human use, we want to limit their use in animal production. They should not be sued solely to help animals gain weight. Yet, we do not want food animals to suffer or become sick. Our governments and the industry are still attempting to discern which antibiotic uses are needed for animal health and which should not be permitted.

Hormone inputs are used to augment animal growth. Governments have based their decisions to allow hormone supplements on scientific studies showing no adverse effects on human health. Uses of beta agonists are regulated so that residues remaining in meat products do not adversely affect human health. The use of hormones and beta agonists reduces amounts of feed needed to produce animal products, thereby reducing the need for cropland, farm machinery, herbicides, and energy.

While pesticide use can harm the environment and pose health risks to humans, governments regulate these inputs so that the risks of harm are less than the benefits accruing from their use. Scientific studies, tolerances, and safety factors are employed in deciding which pesticide uses are permitted. Animal products generally do not contain dangerous pesticide residues. However, vigilance is required to ascertain that the Environmental Protection Agency takes appropriate actions to protect human health. In 2018, groups had to sue the agency to have tolerances for chlorpyrifos residues revoked due to scientific studies showing risks above the safety standards set by the Federal Food, Drug, and Cosmetic Act (*League of United Latin American Citizens vs. Wheeler*, 2018).

Selective breeding, cloning, and genetic engineering have contributed to the more efficient production of animal products. These technologies allow us to produce more products using less feed and energy. In turn, this reduces the carbon footprint of animal production. Yet, potential adverse effects from some of these technologies merit caution.

Suggestions to address inputs

Some consumers feel that uses of antibiotics, hormones, beta agonists, and pesticides are adversely affecting human health. Because these are very different inputs, our regulatory responses will diverge and must be based on individual merits. Most of these inputs reduce costs, reduce energy usage, and are important in alleviating food insecurity.

Reduce uses of antibiotics in animal production

Production agriculture uses large quantities of antibiotics, and some producers administer antibiotics to animals in situations where usage cannot be justified. Because we want to reduce the development of antibiotic-resistant bacteria, further reductions in antibiotic usage in the production of animals are warranted.

Use existing markets for limiting production inputs

After careful study, our federal government allowed the use of hormones, beta agonists, and pesticides in the production of animals under regulatory controls that assure us that animal products are safe. Animal products from animals that did not receive these inputs can be labeled so consumers have the choice of avoiding the inputs. This includes certification programs guaranteeing no hormones or beta agonists were used in the production of animals and the organic label.

Facilitate production to help alleviate food security

There are more than 40 million Americans who are food insecure. They lack sufficient amounts of nutritious food for maintaining a normal life. This reduces the ability of these people to make meaningful contributions through employment. Food insecurity negatively affects our economy.

One of the justifications for using hormones, beta agonists, and pesticide inputs is to reduce prices of foods so fewer people are food insecure. Similarly, genetically engineered plants and animals are important in providing quantities of commodities that help reduce food prices. Efforts to ban these inputs and genetically engineered organisms will increase the numbers of food-insecure Americans.

Marketing and social issues

The marketing of animal products involves labels on products so they can be distinguished from others. Consumers seek information on features accompanying the production of the animals supplying the products. Labels enable consumers to express their preferences for products that are special, which are normally more expensive to produce. Producers will alter production practices and produce more costly products when they are able to recoup their costs through higher prices.

Some consumers want organic products. A defined organic certification program offers a guarantee that producers desisted from using prohibited inputs or practices. There are also markets for locally grown products, and these are popular with many consumers. There is no definition of what is local, so consumers need to inquire about the sources of local products. Purchases of locally grown products support local economies.

Animal waste continues to create problems. Nutrients from animal manure are entering surface waters and denigrating water quality. Although large concentrated animal feeding operations are regulated, too many nutrients from animal waste are entering our surface waters.

Nuisances are another troublesome issue for the producers of our food animals. The smells from large operations and from the application of manure to fields can create objectionable odors and unhealthy conditions for neighbors. While nuisance law provides remedies for unfair situations, every state has enacted an anti-nuisance law that provides a defense for qualifying production facilities. Some states have enacted laws that allow producers to expand exponentially and construct new barns within feet of existing residential neighbors. These laws are unfair.

A few firms producing, transporting, and slaughtering animals have experienced financially detrimental situations due to unauthorized public releases of bad conditions and practices occurring at their facilities. To thwart the release of unauthorized pictures, videos, and audio recordings, several state legislatures enacted "ag-gag" laws to criminalize these activities (Idaho Code, 2018; Iowa Code, 2018; North Carolina General Statutes, 2018; Missouri Annotated Statutes, 2018; Utah Code, 2013). These laws are unnecessary and shield unhealthy and abusive practices. They also may violate the First Amendment freedom of speech (*Animal Legal Defense Fund vs. Herbert*, 2017; *Animal Legal Defense Fund vs. Wasden*, 2018).

Suggestions for responding to marketing and social issues

Four suggestions are offered to support a viable animal production industry. The first lends support to marketers for differentiating animal products. The other three recognize the need for the industry to be more considerate of the rights of others.

Expand consumer choices

Consumers want more information on the attributes of the animals providing food products. To facilitate the differentiation of products, governments can enact laws that define terms to meaningfully distinguish product attributes. State governments might do more to facilitate firms that desire to sell specialized animal products by adopting regulations defining terms to accurately describe selected attributes. This would allow consumers to identify the specialized products they want.

Encourage sustainable practices for handling animal waste

Due to a court order in 1989, the Environmental Protection Agency had to enact more stringent regulations governing animal waste pollution of surface waters. However, more needs to be done to intercept pollutants from our farms so they do not end up in surface waters and the air. If we really want cleaner water and air resources, states need to become more proactive. For some situations, states need more stringent pollution controls. Governments also need to enforce regulations against producers who are flaunting existing regulatory requirements.

Safeguard property rights of neighbors

Some state legislatures have enacted one-sided anti-nuisance laws that declare agricultural production is more important than other land uses (Indiana Code, 2018). These laws completely overrule the rights of neighboring property owners so that a production facility can subject neighbors to terrible conditions. By taking away the nuisance rights of neighbors, anti-nuisance defenses contradict majority rule.

The purpose of nuisance law, an equitable remedy, is to reconcile clashing property rights. Neighbors adjacent to animal production facilities have a right to a healthy environment and deserve to be able to use their properties for reasonable activities. One-sided anti-nuisance laws unfairly obliterate the rights of neighbors, are inimical to public health, and are contrary to community well-being. Changes in these laws are needed to safeguard rights of neighbors who happen to live next to large animal feeding operations.

Support efforts that expose animal abuse and bad practices

Production and marketing firms have received bad publicity due to persons releasing unauthorized pictures and recordings of their facilities. Most of the releases violated existing laws on conversion, trespass, and fraud. There is no need for "ag-gag" laws to provide additional remedies for prosecuting persons involved in these illegal activities. Because the laws curb investigative reporting and shield abusive practices, they should be repealed.

Foodwashing facts

1   Federal agencies are not adequately protecting consumers from fraudulent and misleading labels on animal products.
2   Federal law on animal product labeling unnecessarily precludes constructive state efforts in protecting consumers from fraud.
3   Governments need to become more proactive in helping consumers secure safe seafood products.
4   State legislatures should be more proactive in protecting water resources from pollution and protecting neighbors from egregious smells from animal operations.

References

Animal Legal Defense Fund vs. Herbert. 2017. 263 F. Supp. 3d 1193 (US District Court, Utah).
Animal Legal Defense Fund vs. Wasden. 2018. 878 F.3d 1184 (US Ninth Circuit Court of Appeals, Seattle).
Idaho Code. 2018. Section 18–7042.
Indiana Code. 2018. Section 32-30-6-9.
Iowa Code. 2018. Section 717A.2.
League of United Latin American Citizens vs. Wheeler. 2018. Case No. 17–71636, US Ninth Circuit Court of Appeals.
MacDonald, J.M., Hoppe, R.A. 2017. *Large Family Farms Continue to Dominate U.S. Agricultural Production.* USDA, Economic Research Service.
Missouri Annotated Statutes. 2018. Section 578.625.
North Carolina General Statutes. 2018. Section 99A-2.
Utah Code Annotated. 2013. Section 76–6–112.

Index